AGRONOMY RESEARCH AND DEVELOPMENTS

WEEDS AND THEIR ECOLOGICAL FUNCTIONS

Agronomy Research and Developments

Additional books in this series can be found on Nova's website under the Series tab.

Additional e-books in this series can be found on Nova's website under the e-book tab.

Agriculture Issues and Policies

Additional books in this series can be found on Nova's website under the Series tab.

Additional e-books in this series can be found on Nova's website under the e-book tab.

AGRONOMY RESEARCH AND DEVELOPMENTS

WEEDS AND THEIR ECOLOGICAL FUNCTIONS

ALIREZA TAAB
EDITOR

nova publishers
New York

For permission to use material from this book please contact us:
Telephone 631-231-7269; Fax 631-231-8175
Web Site: http://www.novapublishers.com

Library of Congress Cataloging-in-Publication Data

ISBN: 978-1-62808-332-3

Published by Nova Science Publishers, Inc. † New York

CONTENTS

PREFACE

Weeds are problematic both in agricultural and nonagricultural areas so that they potentially cause economic losses. Weed interference in crops also date back to the beginning of agriculture and they have been able to persist, in spite of long term control operations. They are able to adapt and survive in the environment due to their specific characteristics. Thus, to control weeds efficiently knowledge of weed biology, in particular their survival mechanisms, ecological responses and interactions, is of important value in weed management systems. "Weeds and their Ecological Functions" provides information on weeds functional behaviors in the cultivations including weed seed dormancy, germination and seedling emergence, examples of beneficial and detrimental interactions between weeds and other organisms, weed behaviors in perennial crops and response of weeds to cover crops. This information is useful for students and weed scientists to analyze some of the functional behavior of weeds and to develop integrated weed management systems.

Nowadays, there is a remarkable variability in environmental factors like temperature and precipitation because of climate changes around the globe. Weed and crop growth are substantially influenced by changes in the environment. Therefore, increased knowledge is needed to develop predictive tools to quantify the consequences of the changes. Moreover, weed control in the early stage of growth is of high importance. Factors regulating seedling emergence characteristics of the weeds in the field need to be well understood.

The aspects of the seed dormancy system such as the effect of maternal environmental factors on primary dormancy, dormancy induction and dynamics, germinability and seedling emergence characteristics of weeds, are reviewed and discussed. Discussion is provided on the possible ecological and agricultural implications of weed seed ecophysiology. In addition, examples of seedling emergence behavior of summer and winter annual weed species are presented.

Information is provided on the presence of some weeds in perennial crops that are not always harmful. Some weeds especially those characterized as perennial or invasive are highly competitive and could cause problems to the crops. However, the presence of some other weeds should be clearly reconsidered regarding their utility as cover species and their beneficial impacts. Such an approach could certainly maintain and enhance the profitability of sustainable agricultural systems like organic or low input agriculture.

Cover cropping is a technique to improve soil fertility and to ensure yield stability. An overview is given on cover cropping typologies and their significance in different agricultural contexts, including their effects on physical and chemical soil characteristics, as well on the biocenotic components of the agroecosystems. The main characteristics of the better-suited species for cover cropping purposes are also reported.

There are possible interactions between weeds and other organisms in the environment. To understand the role of weeds in agroecosystems, weed interactions with other biotic compartments including nematodes, plant pathogens, arthropod communities, and a crop are also presented.

Chapter 1 – Weeds have a major impact on crop production by reducing crop quantity and quality. They are also problematic in non-agricultural areas. Therefore, it is necessary to control weeds where they cause a problem. To control weeds efficiently, information on weed biology especially seed dormancy, germination and emergence timing is of vital importance. Dormancy as a seed characteristic determines the range of conditions for germination. A mother plant usually produces seeds with different levels of dormancy predicting uncertainty in the environment for establishment and reproduction among siblings. Seeds with different levels of dormancy demand different conditions to overcome dormancy. This characteristic causes distribution of germination and consequently seedling emergence timing over an extended period of time, which contribute to the seriousness of weed species in the field. Weed control over an extended period is almost impossible due to the crops growing conditions, limitation of applying some weed control measures during advanced crop developmental stages and economic reasons. This may provides opportunity for some weeds to escape environmental hazardous or weed control operations and reproduce even under an efficient weed management program. Therefore, knowledge of weed seed dormancy and germination characteristics and their interaction with environmental factors is important to predict weed seedling emergence timing and maximize the efficiency of weed management strategies.

Chapter 2 – *Carthamus oxyacantha* and *Vaccaria pyramidata* are winter annual weed species frequently found in rainfed crops e.g. wheat and barley. Seedling emergence patterns of *C. oxyacantha* and *V. pyramidata* were studied with seeds buried at soil depths of 0, 2, 4, 6, 8, and 10 cm in outdoor pots during two years. The seedling emergence of *C. oxyacantha* mainly took place during February to March after which it was negligible and sporadic. The seedling emergence slightly delayed with increasing of soil depth. Start of seedling emergence was coincided with a daily mean soil temperature of 5°C at all soil depths. The main flush of seedling emerged when the minimum and maximum soil temperatures were between 4°C to 20°C, respectively. In the first year of the experiment, total seedling emergence was 45.5 % for seeds placed at soil surface, it increased in seeds buried at 2 cm depth (88 %) followed by a decrease at the rest of the depths. However, total seedling emergence did not show a consistent pattern in the second year. The seedling emergence of *V. pyramidata* was also took place from February to March with fewer and sporadic emergences after this time. A minimum daily soil temperature of ca 4°C was found necessary for start of seedling emergence. Most of the seedlings emerged when the minimum and maximum soil temperature were between ca 3 and 14°C, respectively. Total seedling emergence was low for seeds placed at the soil surface in both years (22% in 2010 and 18.5% in 2011); and it was increased at the depth of 2 cm (% 59.5 in 2010 and 51.3% in 2011) followed by a decrease at rest of the depths in 2010. Although large numbers of seedlings of both species may emerge over a short period of time as a main seedling flushes, the sporadic seedling emergences over a relatively extended period enable them to escape weed control operation and maintain problematic in the cultivation.

Chapter 3 – *Chenopodium album* L. is a problematic weed in many crops. Information on seedling emergence characteristics is necessary to optimize the timing of weed control operations. Seedling emergence pattern was studied during two years with seeds of *C. album* buried at soil depths of 0, 2, 4, 6, 8, and 10 cm in outdoor pots. According to the measurements, the mean daily soil temperature was similar at all depths, whereas fluctuation in temperature was higher at the soil surface and decreased with depths. Seedlings emerged from all depths over extended period of time from mid February to mid May. Total numbers of emerged seedlings was higher for seeds buried at the soil surface, possibly due to the stimulatory effect of light and daily fluctuation in temperature on germination, and decreased with depth. However, seedling emergence surprisingly increased again in seeds buried at the depth of 10 cm with a main flush in late March and few sporadic emergences before and after this time. This information can be used to predict seedling emergence behavior in different soil depths and therefore use in management of *C. album* in agriculture.

Chapter 4 – The presence of weeds in perennial crops (orchards, olives, vineyards) is certainly not always harmful.

Obviously, this effect depends on the kind of the present weeds and their relative abundance. Indeed, some weeds (especially those characterized as perennial or invasive) are highly competitive while they could cause several other problems to the crops.

However, the presence of some weeds, especially members of Fabaceae or Poaceae and some other exceptional species (e.g. *Oxalis pes-caprae* L.) should be clearly reconsidered regarding their utility as cover species. Some of these weeds have beneficial impacts by causing soil enrichment and stabilization of semi-stable areas and orchards or being valuable forage.

Moreover, in addition to competition-based weed suppression; certain legumes used as cover crops (hairy vetch, red clover, sweetclover etc.) are also known to suppress weeds through their allelopathic activity both during growth and after their death. Several studies have been conducted in order to evaluate the efficacy of different management systems involving the use of natural vegetation, with respect to the tillage systems generally used in semiarid environments. Future research should focus on legumes or other herbaceous species which can suppress weeds, won't compete for water and nutrients, increase N and organic matter content and control erosion. Such an approach could certainly maintain and enhance the profitability of these sustainable agriculture systems, especially in the case of organic or low input agriculture.

Chapter 5 – Cover cropping has been a known technique since ancient times and has been used in civilizations that are distant from each other, in both space and time, as a tool to improve soil fertility and to ensure yield stability. Over recent years, the growing side effects of intensive agriculture have stimulated a renewed attention toward cover crops, because of their ability to reduce the environmental impact deriving from the increased use of agrochemicals and agricultural mechanization. This chapter provides an overview of cover cropping typologies and their significance in different agricultural contexts, including their effects on physical (structure, porosity, thermal regime, water evaporation, erodibility) and chemical (organic matter content, nitrogen, phosphorus, potassium and microelements) soil characterristics, as well on the biocenotic components of the agroecosystems (weeds, pathogens, arthropods, nematodes). The main characteristics of the better-suited species for cover cropping purposes are also reported. Among these characteristics, particular attention has been paid to soil and climate preferences, vegetative vigour, persistence over time or competitive ability against weeds, all traits considered of pivotal importance in determining their possible diffusion as cover crops in agricultural systems. Finally, it is worth considering that cover crops performance improvement is critically affected by agricultural (i.e. adopting measures aimed at cost reduction, identification of seed mixtures suitable for enhancing the biological complementarity among species) and genetic innovation (breeding of specific genotypes for cover cropping purposes).

Chapter 6 – Although weeds are commonly regarded as being negative influences on crop production, they also contribute toward vegetational diversity in agroecosystems which generally promotes food web stability. Di- and polyculture crop systems have been shown to reduce populations and injury inflected by many crop pests, and weeds have been shown to induce similar effects. Sugarcane, being a perennial plant, offers an illustrative example for the effects of weed communities in and around the field, and sugarcane is arguably one of the most widely studied crops in terms of its relationships with weeds. This review examines what is known about weed interactions with other biotic compartments of sugarcane agroecosystems, including nematodes, plant pathogens, arthropod communities, and the crop itself. Three economically important stalkboring lepidopteran pests of sugarcane, in particular, have received the most attention: the sugarcane borer (*Diatraea saccharalis* [F.]), and the Mexican rice borer, *Eoreuma loftini* (Dyar), the eldana borer (*Eldana saccharina* Walker). *Diatraea saccharalis* interactions with weeds involve enhanced populations of predators, especially the red imported fire ant, *Solenopsis invicta* Buren. *Eoreuma loftini* and *E. saccharina* associations with weeds and other crops occur more on an agricultural landscape scale, wherein plant diversification can exacerbate the spread and degree of infestation. In terms of sugarcane protection, weeds are detrimental as competitors with the crop and as reservoirs for diseases, arthropod pests (particularly during winter), and vectors of sugarcane diseases. In instances where alternate plant hosts of sugarcane pests are problematic, area-wide strategies that account for the effects of alternate host plants on pest populations should be developed to minimize crop injury. It will be necessary to consider the interactions of weeds with sugarcane and its pests on a regional basis in order to address interactions peculiar to each region.

In: Weeds and Their Ecological Functions
Editor: Alireza Taab
ISBN: 978-1-62808-332-3

Chapter 1

WEED SEED DORMANCY, GERMINATION AND SEEDLING EMERGENCE CHARACTERISTICS

*Alireza Taab**
Department of Agronomy and Plant Breeding,
Faculty of Agriculture, Ilam University, Ilam, Iran

ABSTRACT

Weeds have a major impact on crop production by reducing crop quantity and quality. They are also problematic in non-agricultural areas. Therefore, it is necessary to control weeds where they cause a problem. To control weeds efficiently, information on weed biology especially seed dormancy, germination and emergence timing is of vital importance. Dormancy as a seed characteristic determines the range of conditions for germination. A mother plant usually produces seeds with different levels of dormancy predicting uncertainty in the environment for establishment and reproduction among siblings. Seeds with different levels of dormancy demand different conditions to overcome dormancy. This characteristic causes distribution of germination and consequently seedling emergence timing over an extended period of time, which contribute to the seriousness of weed species in the field. Weed control over an extended

* Correspondence: Alireza Taab, Ph.D., Department of Agronomy and Plant Breeding, Faculty of Agriculture, Ilam University, 69315-516 Ilam, Iran. Tel: (+98) 918 6149226; Fax: (+98) 841 2227015; E-mail: a.taab@mail.ilam.ac.ir.

period is almost impossible due to the crops growing conditions, limitation of applying some weed control measures during advanced crop developmental stages and economic reasons. This may provides opportunity for some weeds to escape environmental hazardous or weed control operations and reproduce even under an efficient weed management program. Therefore, knowledge of weed seed dormancy and germination characteristics and their interaction with environmental factors is important to predict weed seedling emergence timing and maximize the efficiency of weed management strategies.

1. Introduction

Weeds cause serious problems in agriculture including reduction of crops yield (e.g. due to interference and competition for resources necessary for plants growth) and crop quality, interfering with harvest and contaminating the harvested products. In addition, some weeds host insects, nematodes, and disease organisms that potentially may attack crops. The negative impacts of weeds on crops can finally lead to a consequent reduction of food production. Therefore, because of increasing demand for food supply, due to the progressively growing human population in the world, weeds interfere in the crop production systems need to be minimized. Weeds interfere in crops date back to the beginning of agriculture. Weeds, however, have been able to escape and survive weed control operations because of their survival mechanisms and adaptability e.g. seed dormancy and germination characteristics. Information on weed biology especially seed dormancy, germination and seedling emergence characteristics can be used to optimize weed management strategies and reduce unnecessary weed control inputs (Grundy, 2003; Juroszek and Gerhards, 2004).

2. Seed Dormancy

Seed dormancy has been defined, in part, differently by different authors (Hilhorst 1995; Dyer, 1995; Baskin andandandandand Baskin, 2004; and Finch-Savage and Leubner-Metzger, 2006). However the most recent and comprehensive explanation of dormancy has been presented by Nikolaeva (2001), Baskin and Baskin (2004) and Finch-Savage and Leubner-Metzger (2006). Dormancy is a seed characteristic that defines the conditions required for its germination, and therefore any factor that widens the range of environmental conditions for germination should be regarded as a dormancy release factor (Finch-Savage and Leubner-Metzger (2006). A dormant seed does not have the capacity to germinate in a specified period of time under any combination of normal physical environmental factors that are otherwise favorable for its germination (Baskin and Baskin, 2004). Seed dormancy and its interactions with the environment, is known to be a highly complex phenomenon (Dyer, 1995). Changes in dormancy could be measured, to somehow indirectly, by setting up germination tests at different conditions for a period of time e.g. 10 to 14 days at constant or fluctuating temperatures in light and darkness (Taab, 2009), under conditions that would normally support germination of that species in the field (Dyer, 1995). To do so, seeds need to be intact and viable (Hilhorst, 1995) and present a degree of dormancy. Therefore, in addition to a germination test to quantify dormancy, a test for seed viability also needs to be done.

There is a complete classification of seed dormancy presented by Baskin and Baskin (2004) among which physiological dormancy is the most important type of dormancy in weedy point of view. Physiological dormancy is a reversal biological mechanism within seed that control its germination interacting with the environment. This type of dormancy enables the species to distribute their seed germination and consequently seedling emergences over both short time e.g. in a season, and long time e.g. for many years.

There is a likely trade off between dormancy and other plant characteristics i.e. dormancy and seed number or size. For example, *Solanum nigrum* L. produces larger number of seeds than *Solanum physalifolium* Rusby. Seed coat in the later species was found to be thicker (Taab, Personal observations). Seeds of *S. nigrum* are less dormant than that of *S. physalifolium* (Taab, 2009). Karssen (1970) also observed that seed coat in dormant seeds of *Chenopodium album* L. was thicker.

2.1. Primary Dormancy

Primary dormancy develops in seeds when they are on the mother plant (Hilhorst and Karssen, 1992). Primary dormancy development is known to be under control of environment experienced by the maternal plant prior to seed dispersal (Fenner, 1991; Hilhorst and Karssen, 1992; Munir *et al.*, 2001) such as temperature (Peters, 1982; Strand, 1989; Fenner, 1991; Roach and Wulff, 1987; Steadman *et al.,* 2004), photoperiod (Munir *et al.*, 2001), age of the mother plant during maturation and position of the seed on the plant (Fenner,

1991), precipitation (Strand, 1989; Philippi, 1993), and soil moisture (Peters, 1982), or the nutritional status of the mother plant (Fawcett and Slife, 1978; Watson and Watson, 1982) during seed maturation. In addition, variation in primary dormancy may be influenced by genes expresses in the maternal parent (Munir et al., 2001), alterations in phytohormone concentrations (Dyer, 1995; Kermode, 2005) and level of abscisic acid (Karssen et al., 1983; Hilhorst and Karssen, 1992) due to environmental factors during seed development.

Dormancy is a complex phenomenon that regulates germination and emergence timing of many weed species interacting with environmental signals depending on type and degree of dormancy and possibly genetic constitution of the species. Gu et al., (2004) reported the presence of genetically complex networks in the regulation of variation for seed dormancy in natural populations of weedy *Oryza sativa* L. *Datura Ferox* L. displays some type of embryo dormancy. Dormancy in *Abutilon Theophrasti* Medik. appeared to be entirely due to its seed coat (Dorado et al., 2009). *Euphorbia esula* L. displays coat-imposed seed dormancy that can be overcome by after-ripening under moist conditions (Foley, 2004). *Aegilops cylindrica* Host. seeds present dormancy that can be, in part, alleviated by after-ripening (e.g. at 22 ± 2 °C) or glume removal, (Fandrich and Mallory-Smith, 2006).

Seed dormancy in *Avena fatua* L. is found to be regulated by three major genes and interaction with environmental factors (Foley and Fennimore, 1998). Leon et al., (2006) showed differences in primary dormancy among biotypes of *Amaranthus tuberculatus* (Moq.) J.D.Sauer in which the rate of seed dormancy alleviation differed between biotypes. Seed protein profiles obtained from the three biotypes differed in protein abundance, number, and type suggesting that differences in seed dormancy levels between the biotypes could be due to different physiological regulatory mechanisms.

Plants grow in different environments may produce seeds with different level of dormancy (Kegode and Pearce, 1998; Koorneef et al., 2002; Steadman et al., 2004; Finch-Savage and Leubner-Metzger, 2006). In addition, there is variation in status of seed dormancy among species (Kon et al., 2007) as well as mother plants, populations and years as shown in *Silene noctiflora* L., *Sinapis arvensis* L., *Spergula arvensis* L., *Thlaspi arvense* L. (Andersson and Milberg, 1998), and among populations of *Avena ludoviciana* and *A. fatua* (Taab and Striebig, 2005), which must be taken into consideration in seed research studies. Toorop et al., (2012) identified three functional groups of seed dormancy, varying in dormancy depth, among genotypes of *Capsella bursa-pastoris* (L.) Medik. with early, intermediate or late flowering time.

They stated that weaker secondary dormancy of early-flowering genotypes may confer a selective advantage in arable fields. Seed populations from plant grown in areas with colder climate appear to have stronger dormancy than those grown in warmer climate (Roach and Wulff 1987; Allen and Meyer, 1998). Seed populations from intermediate climate exhibit variation in dormancy levels, both among and within plants, which enable them to spreads germination across a considerable time periods (Allen and Meyer, 1998). Seeds of *Lolium rigidum* L. (Steadman et al., 2004) and *Goodenia fascicularis* F.Muell. and Tate (Hoyle et al., 2008) from plants grown at warm temperatures were less dormant than seeds from a cool environment. In contrast, seeds of *Setaria faberi* Herrm. raised under lower temperatures were less dormant than those that ripened under higher temperatures (Kegode and Pearce, 1998). Seeds of lettuce produced under higher temperatures (30/20°C vs. 20/10°C) had lower level of dormancy e.g. thermoinhibition and photodormancy (lack of germination in darkness), linked to reduction in abscisic acid (ABA) content in seeds (Contreras, 2007).

Figueroa et al., (2010a) studied seeds populations of *Senecio vulgaris* L. occurring at six sites along a 700-km north–south transect (Kentucky to Michigan) in the USA. They observed two germination patterns. Seeds from the southern locations averaged 80 to 90% germination across the range of 5 to 25 °C; however, seeds from northern locations had reduced germination when incubation temperatures were close to 5 or 25 °C. The test with their progeny seeds from all locations grown in a common environment, suggested that germination of the parent seeds could be due to maternal environmental effects. Figueroa et al., (2010b) showed that seeds from plants of *S. vulgaris* grown under warm environments produce non-dormant seeds. In contrast, more dormant seeds produced by plant grown in cool environment, suggested that dormancy can be induced by the cool maternal environment. Yamasue et al., (1993) concluded that the strength of seed dormancy in *Echinochloa oryzicola* (Vasinger) Ohwi increases with latitude in the northern hemisphere.

Photoperiod and light quality in the environment during seed maturation was found to affect the primary dormancy (Hilhorst and Karssen, 1992; Armstrong and Adkins, 1998; Munir et al., 2001; Contreras, 2007). Plant populations of *A. fatua* grown under short photoperiods (from 18h to 10h) produced fewer seeds with a higher degree of dormancy (Armstrong and Adkins, 1998). Plants of *C. album* grown in a continuous 16 h photoperiod produced more dormant seeds than plants grown in a continuous 8 h photoperiod (Karssen, 1970). The seeds of *C. album* ripened in long days (August) revealed a strong primary dormancy. However, the seeds ripened in short days (October) were partially dormant after the harvest (Jursík et al., 2003). The strongest primary dormancy was also found with the seeds of *Galinsoga ciliata* (Raf.) Blake ripened in long days (in July and August). A shorter length of the day led to a weaker primary dormancy (Jursík et al., 2003). *Arabidopsis thaliana* (L.) Heynh. seeds matured by parents during short days break dormancy more readily when stratified, simulating spring condition; but seeds matured during long days break dormancy more readily on average when not stratified, simulating autumn conditions. Therefore, the photoperiod during seed maturation might be a reliable indicator of season (Munir et al., 2001). Seeds collected from *A. thaliana* plants grown under short days were also more deeply dormant than that of grown under long days (Dasti et al., 2001).

Seeds of lettuce (*Lactuca sativa* L.) produced under higher red: far-red (R:FR) ratios had higher germinability, lower sensitivity to external ABA, and lower ABA concentrations indicating higher accumulation of the active form of phytochrome (Pfr) in seeds drying under red-rich light. This can be a reason for production of lettuce seed with lower levels of thermoinhibition and photodormancy (Contreras, 2007). Light quality influences the proportion of phytochrome that will be stored in the dry seeds in the form of the active far-red absorbing from (Pfr) or one of its intermediates that can easily be converted to Pfr in darkness upon imbibition. Parallel to other processes in plants the effects of temperature and day length might be linked to hormone levels and hormone sensitivity (Hilhorst and Karssen, 1992).

Growing conditions during plant life cycle can affect dormancy status in mature seeds (Brainard et al., 2005; Clua et al., 2006; Jha et al., 2010; Williams II et al., 2012). Brainard et al., (2005) found that dormancy was higher in seeds of *Amaranthus powellii* S.Wats. maturing on plants grown under competition than without competition. They concluded that increased seed dormancy may be an adaptive response to canopy shade. In *Amaranthus palmeri* S.Wats., increasing the shading increased the proportion of dormant and ABA content of seeds. Seeds that matured in the top and middle third of the mother plant were less dormant than those that matured in the bottom third of the mother plant. Endogenous gibberellic acid (GA) content of seeds did not differ between locations on the mother plant; however, the ABA content of seeds produced on the bottom third of the plant was higher than the ABA content of seeds produced at the middle and top third of the plant (Jha et al., 2010). The size and germinability of *Panicum miliaceum* L. seeds was highest in seeds produced in less-competitive maternal corn environments (Williams II et al., 2012). Plants of *A. theophrasti* growing in relatively noncompetitive environments or that germinate early in the season are likely not only to produce a greater number of seeds but also a greater proportion of seeds that are dormant (Nurse and DiTommaso, 2005). Low water availability in drought stress enhanced the number of dormant seeds in *Lotus glaber* Mill. (Clua et al., 2006).

Disturbance, farming system and crop rotation history may affect dormancy formation in seed populations. Brainard et al., (2007) suggested that perennial alfalfa rotations may select for greater dormancy and longevity of *A. Powellii* seeds. Brainard et al., (2006) showed that both seed germination and emergence in *A. powellii* were greater for seeds originating from organic compared with conventional vegetable farms, indicating lower dormancy in seeds from organic farms. Some *Bromus diandrus* Roth. populations from cropping fields showed longer seed dormancy than those collected from adjacent non-cropped fencelines (Kleemann and Gill (2012). Dormancy was also stronger in populations of *Hordeum murinum* ssp. *glaucum* (Steud.) Tzvelev. collected from cropping fields than populations from non-crop habitats (Fleet and Gill (2012). Thus, stronger dormancy state in weed seeds could be an adaptation strategy in response to disturbances in the habitats. It can be speculated that every disturbance regime may act as a selection pressure (Dekker J., Personal communication) so that weeds react to it by adapting their dormancy and germinability and consequently regulate their emergence timing to survive and maintain in the environment. Seed dormancy, therefore, is seen as a survival mechanism.

2.2. Dormancy Reduction

Seed dormancy can be reduced during stratification ((Finkelstein *et al.*, 2008; Taab and Andersson, 2009b; Kleemann and Gill, 2012; Fleet and Gill, 2012) and/or dry after ripening (Jursík et al., 2003; Fleet and Gill, 2012). Dormancy can also be reduced by overwintering (Taab and Andersson, 2009a; Dorado et al., 2009; Schutte et al., (2012) similar to cold stratification, characteristics of summer annual species (Baskin and Baskin 1988; Baskin and Baskin, 2004). In contrary, cold stratification may cause dormancy induction in winter annual species (Milberg and Andersson 1998). The stratification environment experienced by seeds can affect the dormancy status or the expression of light requirement for germination (Noronha et al., 1997).

Dormancy reduction by stratification has been shown in seeds of many weed species e.g. *S. nigrum* (Taab and Andersson, 2009b), *B. diandrus* Roth

(Kleemann and Gill, 2012), *H. murinum* ssp. *glaucum* (Fleet and Gill, 2012). Dry condition was found to be influential in breaking dormancy in wild rice seeds, whereas moist condition was unfavorable (Hyakutake and Yoshida, 1993). Dry storage at 5°C for 6 months and at 20/30°C for 3 months did not reduce dormancy in seeds of *S. physalifolium* (Bithell et al., 2002).

Increased rate of dormancy release with stratification temperature has been shown in *S. nigrum* seeds (Roberts and Locket, 1978; Taab and Andersson, 2009b; Bithell et al., (2002). Cold stratification is found to be the most effective factor for breaking seed dormancy in *Solanum sarrachoides* Sendtn.. Moist compared with dry storage was more effective at 4ºC, but dry storage was more effective at 17ºC (Zhou et al., 2005a). Come and Spruit (1983) concluded that 10 days of stratification at 2°C is needed for loss of dormancy in *A. thaliana*, whereas 1 or 2 days at higher temperatures e.g. 15°C, was enough to release dormancy.

Cold stratification at 4ºC was found to be effective for releasing dormancy in seeds of *Impatiens glandulifera* Royle (Mumford 1988), *S. sarrachoides* (Zhou et al., 2005a) and in *Amaranthus tuberculatus* (Leon and Owen, 2003). The process of dormancy breakage in *I. glandulifera* can be interrupted by a period at 20°C (Mumford 1988). Willemsen (1975) reported that stratification at 4°C is the most successful treatment for breaking dormancy in seeds of *Ambrosia artemisiifolia* L., whereas -5°C is least effective and 10°C is intermediate. Brändel (2005) found out primary dormancy in *Carex pendula* Huds. can be broken by stratification temperatures of 5, 11, 13°C but 5°C was the most effective. For seeds of *C. remota* primary dormancy could be relieved under stratification at 5°C. Seed dormancy in *Najas marina* L. can be broken at temperatures below 15°C (Handley and Davy, 2005). Steadman (2004) reported that dormancy release of *L. rigidum* seeds stratified at 10 °C and 30/10 °C reached a maximum after 40 and 20 days, respectively, and no further improvement in germination was gained by extending stratification time. Dormancy can be reduced in *H. murinum* ssp. *glaucum* during dry after-ripening or stratification. *H. murinum* ssp. *glaucum* populations from cropping fields and noncrop habitats showed large differences in their pattern of dormancy loss. Seeds from noncrop habitats may loss their dormancy readily under dry storage. In contrast, seeds from cropping fields are found to be more responsive to cold stratification to loss dormancy (Fleet and Gill, 2012). Loss of dormancy in seeds of *D. ferox* can be achieved by exposure of the seeds to natural chilling conditions (Dorado et al., 2009). Schutte et al., (2012) showed that seed dormancy in *Ambrosia trifida* L., including embryo and embryo-covering structure-enforced dormancy, can sequentially be reduced by overwintering. Dormancy release has been reported due to after-ripening (kept in a dark, dry place at 20°C) in seeds of *C. album* and *G. ciliata*. The length of primary dormancy varied from 10 to 100 days depending on the date of seed ripening and on the year (Jursík et al., 2003). In general, winter annual species are more responsive to dryness after ripening to release dormancy, simulating the summer condition before seedling emergences. In contrast, summer annual species are more responsive to moist cold stratification to release dormancy, simulating autumn and winter conditions before seedling emergences.

2.3. Secondary Dormancy

Seeds with released primary dormancy may gradually enter the state of dormancy again if the proper conditions for their germination e.g. temperatures, light, oxygen and moisture (Baskin and Baskin 1998; Foley, 2001) are not present. Non-dormant seeds exposed to high or low temperatures for a period of time, darkness, buried deep in the soil, faced to lack of O_2 or to high concentration of carbon dioxide (Holm 1991; Khan and Karssen, 1980; Dyer, 1995; Benvenuti and Macchia 1995, 1998; Benvenuti et al., 2001; Brändel and Jensen, 2005; Finch-Savage and Leubner-Metzger, 2006) may also enter the state of dormancy. This is called secondary dormancy, which is part of the process of dormancy dynamics in seeds with physiological dormancy.

2.4. Dormancy Dynamics

A cyclic change in levels of seed dormancy has been shown in many species (Esashi et al., 1983; Hilhorst et al., 1986; Taab and Andersson, 2009b) indicating presence of physiological dormancy. Dormancy release and induction have been introduced as two different processes. Both may overlap but due to a different temperature optimum the process of dormancy release may predominate at lower, and induction at higher temperatures (Totterdell and Roberts, 1979; Hilhorst et al., 1986). Moreover, Esashi et al., (1983) hypothesized two counteracting systems for seed dormancy in *Xanthium pennsylvanicum* Wallr. including dormancy breakage and dormancy inducing systems. Dormancy levels generally may cycle between dormancy or conditional dormancy and non-dormancy depending on the species (Baskin and Baskin, 1998) in the nature. It is concluded that dormancy is a dynamic process that may change gradually between highest to lowest levels and vice versa, depending upon environmental conditions they are likely to experience.

Rate of dormancy reduction was shown to be low in seeds of *S. nigrum* stratified at 4.5°C but it was increased at 10 and 15.2°C. An initial dormancy reduction followed by induction was observed at 15.2 and 18.6°C (Taab and Andersson, 2009b). A similar result was also observed in seeds of *S. physalifolum*. A rapid initial reduction followed by induction of dormancy was also observed in seeds of *S. nigrum* and *S. physalifolum* stratified at temperatures of 18, 20, 22, and 24ºC (unpublished data). This dormancy response to temperature was also confirmed by a test at weekly stepwise increasing temperatures of 4.5, 10, 15.2, 18.6, 25.4, 30.2, 35.2 and 40ºC for 8 weeks (Taab and Andersson, 2009b). Hilhorst et al., (1986) showed that a pretreatment stratification at 2°C is required to weaken seed dormancy in Sisymbrium officinale (L.) Scop. They concluded that dormancy could be released at a much wider range of temperatures. At higher temperatures the releasing of primary dormancy occurred at a much higher rate. However, the same is true for the induction of secondary dormancy. Therefore, the period of optimal dormancy release occur markedly earlier at higher temperatures but also last a much shorter time.

Seeds of *X. pennsylvanicum* showed a dormancy that could be broken by stratification at 8 or 23°C. Prolonged stratification at the same temperatures caused induction of secondary dormancy (Esashi et al., 1983). Totterdell and Roberts, (1979) showed that loss of seed dormancy in *Rumex crispus* L. and *R. obtusifolius* L. can occur over a wide range of stratification temperatures, from 1.5 to 15°C. But 1.5°C was the most effective and the optimum period of stratification at any given temperature decreased with increase in temperature. There was no effect on dormancy release at 20°C. They concluded that in addition to loss of primary dormancy, an induction of secondary dormancy can also occur during stratification. They indicated that the critical temperature below which dormancy is rapidly lost, must be between 15 and 20°C. Seed dormancy in *Najas marina* can be broken at temperatures below 15°C and the germinability can reach a peak after 10-12 weeks of stratification at 4°C (Handley and Davy, 2005). Brändel and Jensen (2005) showed that primary dormancy in *Eupatorium cannabinum* L. achenes could be relieved by stratification temperature. With prolonged period, certain stratification temperatures led to a more or less continuous increased dormancy release. They also revealed significant effects of stratification temperature, stratification period, germination test temperature and all interactions between these factors. As a wetland plant, primary dormancy in *Eupatorium* achenes was relived at temperatures between 3 and 18ºC. Consequently even higher temperatures can lead to a loss of dormancy in *Eupatorium* achenes. However, in *Eupatorium* achenes low stratification temperatures were more effective in relieving primary dormancy than high temperatures. On the other hand, an increase of temperature during stratification for certain period led to induction of dormancy.

2.5. Seasonal Dormancy Cycle

Seasonal changes in seed dormancy, expressed in germinability or emergence pattern, have been shown in many species e.g. in *S. nigrum* (Roberts and Lockett, 1978; Taab and Andersson, 2009a), *Lamium amplexicaule* (Baskin and Baskin, 1981), *Galeopsis speciosa* Mill., *Polygonum lapathifolium* L., *Polygonum convolvulus* L., *Matricaria perforata* Merat, *Thlaspi arvense* L., *Stellaria media* (L.) Vill. and *Capsella bursa-pastoris* (L.) Medik. (Håkansson, 1983), *S. sarrachoides* (Roberts and Boddrell, 1983), *Alopecurus myosuroides* Huds., *Bromus sterilis*, *Papaver rhoeas*, *Veronica arvensis*, *Veronica persica and Viola arvensis* Murr. (Froud-Williams et al., 1984), *Polygonum persicaria* L. and *C. album* (Bouwmeester, 1990, Bouwmeester and Karssen, 1992 and 1993; Vleeshouwers and Kropff, 2000; Vleeshouwers and Bouwmeester, 2001), *Monochoria vaginalis* (Chen and Kuo, 1999), *Lesquerella fendleri* (Hyatt et al., 1999), *Echinochloa crus-galli* (Honek et al., 1999*)*, *Spergula arvensis* (Vleeshouwers, 1997ab; Vleeshouwers and Kropff, 2000; Vleeshouwers and Bouwmeester, 2001), *Poa annua* (Shem-Tov and Fennimore, 2003), *Galium aparine* and *Brassica kaber* (Mennan and Ngouaji 2006), *Papaver rhoeas* (Cirujeda et al., 2006) and *S. physalifolium* (Taab and Andersson, 2009a). Seeds may undergo seasonal dormancy cycling if conditions are suboptimal e.g. unfavorable temperature conditions or lack of adequate light or nitrate for their germination, progressively gaining or losing dormancy until they eventually germinate or die (Finkelstein *et al.*, 2008). Cyclic changes in dormancy are regulated by endogenous rhythms and seasonal changes of temperature (Froud-Williams et al., 1986). Dekker (2003) stated that after-ripening, the occurrence and timing of seedling emergence, and the induction of secondary dormancy are regulated by seasonally and diurnally varying soil oxygen, water, and temperature signals in soil seed pools.

Temperature is known as the main factor regulating dormancy changes (Bouwmeester, 1990; Hilhorst and Karssen, 1992; Bouwmeester and Karssen,

1992, 1993; Baskin and Baskin, 1998; Benech-Arnold et al., 2000 Taab and and Andersson, 2009a). Other factors are not considered to affect dormancy directly e.g. soil moisture (Bouwmeester, 1990; Bouwmeester and Karssen, 1992), nitrate (Bouwmeester, 1990; Bouwmeester andd Karssen, 1992, 1993), light and desiccation (Bouwmeester and Karssen, 1993). However, some authors stated that fluctuations in soil moisture (e.g. in *Polygonum aviculare* L.) under field conditions (Batlla and Benech-Arnold, 2006), and a hydration-dehydration cycle (e.g. in *Polygonum lapathifolium subsp. Lapathifolium* L.) under laboratory conditions (Bouwmeester, 1990) or desiccation may relieve dormancy (Vleeshouwers, 1997a).

Therefore, dormancy cycle is mainly related to temperature changes (Hilhorst and Karssen, 1992) and associated with changes in the range of environmental conditions in which seeds are able to germinate (Vleeshouwers and Kropff, 2000). In summer annual species, stratification under lower temperature in the soil during winter weakens dormancy thus enabling seeds to germinate in spring. However, high temperatures during summer cause dormancy induction that prevents seed germination in late summer or early autumn (Baskin and Baskin, 1998) as shown in e.g. *S. nigrum* and *S. physalifolium* (Taab and Andersson, 2009a). In contrast to summer annuals, dormancy becomes stronger in winter annual species after cold stratification and is released at high temperatures (Milberg and Andersson, 1998) as shown in e.g. *Apera spica-venti* (L.) P.Beauv. and *Alopecurus myosuroides* (Andersson and Åkerblom-Espeby, 2009). Generally, dormancy is reduced during autumn, winter and spring and induced during summer in summer annual species. On the other hand, in winter annual species dormancy is reduced during summer and could be induced during autumn, winter and spring. However, some species may show both winter annual and summer annual behavior. As shown in *Arabidobsis thaliana* (Derkx and Karssen, 1994; Munir et al., 2001) a main flush of seedlings emerges in autumn and a small flush emerges in spring.

3. Germination

Seeds only germinate when they are non-dormant and factors required for their germination are present. Environmental factors including temperature, light, nitrate, desiccation (Hilhorst and Karssen, 1992; Bouwmeester and Karssen, 1993) and moisture affect seed germination. Finkelstein et al., (2008) defined germination as the initial emergence of the radicle from the seed coat.

In the presence of environmental factors, germination process requires that the plant embryo leaves the quiescent state, mobilizes stored nutrients, overcomes the barrier of surrounding tissues, and resumes cell elongation, cell division, and development. Dormancy may result from blocks in any of these processes (Finkelstein *et al.*, 2008) depending on type of dormancy.

Seed dormancy and germination process are sometimes confounded with each other because of interactions between factors regulating both processes, complexity of these systems and differences among species, e.g. some factors like light or fluctuating temperatures may terminate germination. Ramon and Owen (2003) reported that seed dormancy of *Setaria faberi* Herrm. is partially phytochrome regulated, but high temperature exerts a more important regulation. Seed dormancy of *Amaranthus tuberculatus* (Moq.) J.D. Sauer. was also found to be regulated by phytochrome. However, high temperatures overcome the inhibitory effect of far red light and promoted the germination of chilled seeds. According to Donohue et al., (2007) phytochrome pathways that mediate germination appear to be crucial sensors not only of light conditions but also of temperature conditions. The pathways that are mediated by phytochrome have major effects on how seeds germinate in response to multiple seasonal cues.

There are likely variations in seed germination characteristics among weed species as well as among populations within species. Li and Feng (2009) observed differences in germination capacity and index among 14 populations of *Eupatorium adenophorum* Spreng. at the same temperature or water treatment indicating genetic differentiations in these traits. Germination capacity and index were also different for each population across temperature or water treatments, showing phenotypic plasticity.

3.1. Factor Affecting Germination

There are several environmental factors affecting seed germination including temperature, light, moisture, salinity, nitrate, gasses, gibberellic acid and pH. There could also be interactions between theses factors.

3.1.1. Temperature

There is a range of temperature e.g. minimum, optimum and maximum, over which seeds usually germinate (Vleeshouwers and Bouwmeester, 2001; Penny and Neal, 2003; Taab and Andersson, 2009b). Germination will start when the temperature rise above minimum/base temperature for germination.

The rate and germination percentages are increased by increase of temperature and reached their maximum at a temperature that is called optimum for germination. Afterward, further increase of temperature may cause reduction in germinability until it becomes zero at a temperature called maximum/upper limit temperature for germination (Table 1). Therefore, there is a functional relationship between temperature and germination. This relationship might be a linear (triangle) or non-linear (bell shaped) function depending on species (Figure 1). Kebreab and Murdoch (1999) stated that the temperature limits vary quite widely and are normally distributed in the seed population. Dormancy release results in a widening of the range of temperatures over which germination can occur, and dormancy induction results in a narrowing of this range (Vleeshouwers and Bouwmeester, 2001). Lowering of the base temperature for germination associated with loss of dormancy has been shown in several species e.g. *S. physalifolium* (Del Monte and Tarquis, 1997), *Aesculus hippocastanum* L. (Steadman and Pritchard, 2003) and *P. aviculare* (Batlla and Benech-Arnold, 2003).

The temperature limits for germination can be examined experimentally by conducting germination tests with non-dormant seeds or seeds with different levels of dormancy at a wide range of constant temperatures. Seeds of *S. nigrum* with a reduced level of dormancy, i.e. stratified at 5 and 15°C for 38 days, germinated at constant temperatures of 18 to 34°C with an optimum temperature from 26 to 30°C (Taab and Andersson, 2009b). The seeds of *G. ciliata* germinated at 10 to 33°C with maximum germination (>85%) at 12–28°C. Seeds of *C. album* germinated in a wider temperature interval (5–33°C). Maximum germination (75%) was obtained at 18°C (Jursík et al., 2003).

Singh and Singh (2009) studying the range of temperature for seed germination, averaged over 12 weed species and temperatures, reported highest weed seed germination in the temperature regime of 25 to 35°C with maximum being at 30°C. The reported temperature limits for germination of some weed species are summarized in Table 1.

There is a relationship between temperature and rate of germination interacting with dormancy level in seeds with physiological dormancy. This relationship is conceptually shown, as a linear response, in Figure 1. However, it should be noted that it can also be a nonlinear response in some species. Seeds with lowest level of dormancy germinate at a wider range of temperatures from lower to upper limit for germination. This range of temperature is called temperature window for germination. The temperature range becomes narrow toward the optimum temperature by induction of dormancy and at highest level of dormancy the germination is completely blocked.

The opposite is true for reduction in dormancy level. When the dormancy becomes weaker the temperature range for germination becomes wider toward its maximum range.

Table 1. The ranges of temperature for germination of some weed species

Species	Temperature °C			Reference
	Base	Optimum	Maximum	
Amaranthus albus	10-12	28-36	-	Sokolova, 2008
A. blitoides	7-8 10	30 – 36 >20	- -	Sokolova, 2008 Steckel et al., 2004
A. retroflexus	6-8 10 -	26-36 >20 -	- - >35/30	Sokolova, 2008 Steckel et al., 2004 Guo and Al-Khatib, 2003
Anthemis arvensis	2-3	-	22-28	Nadtochji, 2008
Avena fatua	5-8	-	20-30	Luneva, 2008
Caperonia palustris	20	30	40	Koger et al., 2004
Capsella bursa-pastoris	1-2	15-26	32-34	Larina, 2008
Cardaria draba	4-5	15-20	30-35	Larina, 2008
Chenopodium album	5 3-4	18 20-30	33 34-36	Jursík et al., 2003 Larina, 2008
Convolvulus arvensis	4-6 10	18-24 20-35	- 40	Nadtochii 2008 Brown and Porter, 1942
Cuscuta campestris	13-14	20-30	-	Doronina, 2008
Eupatorium adenophorum	10	25	30-35	Lu et al., 2006
E. cannabinum	3	27-30	38	Brändel and Jensen, 2005
Fallopia convolvulus	2-4	14 - 16	35	Doronina, 2008
Flaveria bidentis	15	22.5 - 35	40	Li et al., 2011
Galinsoga ciliata	10	12-28	33	Jursík et al., 2003
Orobanche cumana	10	25	35	Yakutkin, 2008
Portulaca oleracea	-	25-35	52-54	Nadtochii 2008
Raphanus raphanistrum	2-4 5	- 20-25	- 35	Larina, 2008 Cheam and Code, 1998
Reseda lutea	2	12 - 20	34	Nadtochii 2008
Senna obtusifolia	15	15-30	50	Norsworthy and Oliveira, 2006
Sinapis arvensis	2-4	14-20	-	Larina, 2008
Sisymbrium altissimum	4	18-20	40	Sokolova, 2008

Table 1. (Continued)

Species	Temperature °C			Reference
	Base	Optimum	Maximum	
Solanum nigrum	18	26-30	34	Taab and Andersson, 2009b
	4	-	30	Roberts and Lockett, 1978
	10	20-30	36	Givelberg et al., 1984; DelMonte and Tarqius, 1997; Larina, 2008
S. physalifolium	12.5	30	-	DelMonte and Tarqius, 1997
S. sarrachoides	19	27-33	39	Zhou et al., 2005a
S. ptycanthum	20	28-33	-	Zhou et al., 2005b
S. melanocerasium	17	22-30	35	Zhou et al., 2006
Thlaspi arvense	2-4	20-24	34-36	Larina, 2008
Xanthium strumarium	14-16	20-24	32-34	Nadtochii 2008

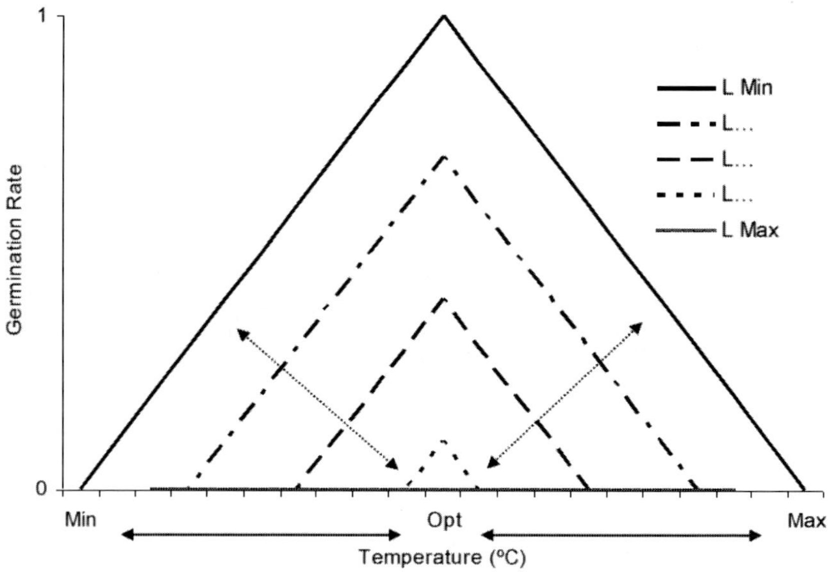

Figure 1. The functional relationship between temperature and germination rates in seeds with different level of dormancy. L Min to L Max: represent possible gradual changes in reversible minimum and maximum dormancy levels expressed in germination; Min, Opt, and Max: represent the minimum, optimum and maximum temperatures for germination.

There is a contradiction in reported base temperature for germination of some species among the literatures (see Table 1). This contradiction could be due to differences between populations in their dormancy, seed handling and treatments during collection, storage and experimental conditions. These all may affect the level of dormancy. For instance, the lower the level of dormancy the lower the base temperature for germination would be. This may affect the preciseness of the computational models to predict seedling emergence of weed species in the field. Thus, it has to be taking into consideration in weed research studies.

Seeds of some weed species germinate at constant temperature. However, alternating temperature is known to promote germination substantially in many species (Bouwmeester and Karssen, 1993; Huarte and Benech-Arnold, 2005; Taab andd Andersson, 2009b). For example, alternating temperature regimes were found to increase total germination of *Amaranthus* species including *A. palmeri*, *A. blitoides* S.Wats., *A. hybridus* L., *A. spinosus* L., *A. retroflexus* L., *A. tuberculatus* (Moq.) J.D.Sauer, and *A. albus* L., except *A. powelli*, which germinated similarly under both constant and alternating temperatures (Steckel et al., 2004). Germination peaked at 25/20 °C in *A. rudis* and at 35/30 °C in *A. retroflexus* and *A. palmeri*. Seed germination of all three species declined when temperatures increased above 35/30 C (Guo and Al-Khatib, 2003). In *Portulaca oleracea* L., germination was lower at 25/15°C and 35/25°C than at 30/20°C (70%, 75% and 81% germination, respectively) in the light/dark regime (Chauhan and Johnson, 2009a). Chejara et al., (2008) studying the effect of diurnal alternating temperatures combinations from 5 to 45 °C with a 12-h photoperiod in *Hyparrhenia hirta* (L.) Stapf found that the speed of seed germination was very high at 30/20 °C (more than 80% of seeds were germinated within 12 h). Alternating temperatures of 17.5/32.5 °C (15 °C daily fluctuation) in *Xanthium strumarium* L. (Norsworthy and Oliveira, 2007), 30/20 °C in *Eclipta prostrata* (Chauhan and Johnson, 2008b), 30/25 and 35/30 °C in *Solanum ptycanthum* Dun. (Zhou et al., 2005), 25/15 °C in *Solanum elaeagnifolium* Cav. (Stanton et al., 2012) were found best for seed germination. The germination of *Raphanus raphanistrum* L. was 1.2 to 1.5 times greater at alternating temperatures (2.5/17.5, 7.5/22.5, and 12.5/27.5 °C) than at constant temperatures (10, 15, and 20 °C) (Malik et al., 2009). Bouwmeester and Karssen (1993) showed that buried seeds of *C. album* do not germinate in darkness at constant temperatures, but alternating temperatures cause considerable germination. Germinability in seeds of *S. nigrum*, tested at alternating temperatures, increased at amplitudes between 5 and 15°C but was reduced at 20°C (Taab and Andersson, 2009). Thus, there seem to be not only an optimum constant temperature but also an optimum alternating temperature for germination.

Sensitivity to temperature fluctuations is known to act as a depth sensing mechanism (Thompson and Grime, 1983). There is a greater temperature fluctuation at the soil surface, and the fluctuation is decreased with increasing of soil depth (Taab et al., 2011ab). Seed response to fluctuating temperature for germination enables the species to detect the appropriate position and time for germination and subsequently time their emergence to the appropriate conditions for seedling establishment and reproduction (Taab, 2009).

3.1.2. Light

Light promote germination in seeds of many species e.g. *S. vulgaris* L. and *C. bursa-pastoris* (Popay and Roberts, 1970), *A. retroflexus* (Schonbeck and Egley, 1980), *C. album* and *G. ciliata* (Jursík et al., 2003), *Fatoua villosa* (Thunb.) Nakai (Penny and Neal, 2003), *E. adenophorum* (Lu et al. 2006), *Sisymbrium orientale* L. (Chauhan et al., 2006c), *P. oleracea* (Zimmerman, 1977; Chauhan and Johnson, 2008c; Chauhan and Johnson, 2009a), *Eclipta prostrata* (L.) L. (Chauhan and Johnson, 2008b), *H. hirta* (Chejara et al., 2008), *S. ptycanthum* (Zhou et al., 2005ab), *S. nigrum* (Taab and Andersson, 2009b), *Flaveria bidentis* Juss. (Li et al., 2011).

Some species may also germinate equally in light and darkness. Seeds of *S. sarrachoides* (Zhou et al., 2005ab) and *S. melanocerasium* (Zhou et al., 2006) germinated equally well under both a 14-h photoperiod and continuous darkness. Seeds of *Cassia occidentalis* (Norsworthy and Oliveira, 2005) and *Solanum rostratum* (Wei et al. 2009) also germinate equally in light and dark. Chauhan et al., (2006e) reported that germination of *Malva parviflora* L. was not influenced by light.

However, light may inhibit germination in some species. Kleemann and Gill (2012) reported strong inhibition of seed germination in *B. diandrus* populations when exposed to light. Light inhibition of germination was also shown in *Phalaris canariensis* seeds (Matus-Cadiz and Hucl, 2005). Germination of *R. raphanistrum* was consistently greater in the dark than in the light. However, this effect was less apparent when germination temperature was increased (Mekenian and Willemsen, 1975). Germination of *Galium tricornutum* was favored by darkness (Chauhan et al., 2006a). Therefore, seed germination response to light could be a species-specific phenomenon. Small-seed species that usually require light for germination are called photoblastic.

Dormancy release causes a gradual increase in sensitivity to light followed by subsequent stimulation of germination (Steadman, 2004). Derkx and Karssen (1994) stated that seasonal dormancy pattern in *A. thaliana* is mainly regulated by change in sensitivity to light. GA (gibberellic acid) sensitivity contributes to this pattern but is not primarily controlling dormancy. The GA requirement was obvious for germination. Nitrate or sensitivity to nitrate did not contribute to the regulation of dormancy and germination of this species.

Light regulation of germination is mediated by photoreceptors called phytochromes that can initiate or stop physiological processes in the seed depending on the characteristics of the light to which it is exposed (Benech-Arnold et al. 2000). Phytochromes translate light signal into biochemical signal subsequently transduced through largely unknown pathways into molecular and physiological changes that modulate growth and development (Casal et al., 1998). Phytochromes have two photo-interconvertible forms and change their form depending on the wavelength of the light to which they are exposed. They are red light-absorbing form (Pr), R, and the far-red light-absorbing form (Pfr), FR, and are synthesized in the Pr form (Casal et al., 1998; Benech-Arnold et al., 2000). The phytochrome form that absorbs red light (R), Pr, with a wavelength of 660 nm is considered biologically inactive. The form that absorbs far-red light (FR), Pfr, with a wavelength of 730 nm is considered the active form that promotes germination. In addition, when the Pfr form is exposed to FR, it changes to the Pr form. In the same way, when the Pr form is exposed to R, it changes to the Pfr form. Therefore, R promotes germination in photoblastic seeds (Benech-Arnold et al. 2000).

Three mode of action of phytochromes, very-low-fluence responses, Low-fluence responses and high-irradiance responses have been considered to define the quantitative relationship between response and predicted levels of the far-red light absorbing form of phytochrome. These differ at the level of the photoreceptor (Casal et al., 1998). Thus, the full moon light may promote the germination of very light-sensitive weed seeds during nighttime soil disturbance (Juroszek and Gerhards, 2004). For seeds to react to light for germination, they need to be in the imbibed state.

3.1.3. Moisture

For germination to take place seeds need to imbibe enough water and activate biochemical process. Response of seeds to water stress and drought is usually evaluated by germination test at a range of osmotic potential. For example, germination in *P. oleracea* decreased from 71% to 10% as osmotic potential decreased from 0 to −0.6 MPa, and was completely inhibited at osmotic potentials lower than −0.6 MPa (Chauhan and Johnson, 2008c). Germination of *H. hirta* was completely inhibited at an osmotic stress of −0.55 MPa or greater and was reduced by 50% at −0.37 MPa (Chejara et al., 2008). Seed germination of *Rapistrum rugosum* decreased as osmotic potential decreased from 0 to −1.0 MPa with some germination (22%) at an osmotic potential of −1.0 MPa (Chauhan et al., 2006d). Germination of *Caperonia palustris* decreased linearly as osmotic potential increased from 0 to −0.8 MPa with nine percent germination at an osmotic potential of −0.8 MPa indicating that *C. palustris* can germinate under moderate water stress conditions (Koger et al., 2004). For *E. adenophorum* seeds, germination was 95% at osmotic potential of 0 MPa and more than 80% at −0.2 MPa, thereafter the germination was decreased by increasing osmotic potential and no germination occurred at a water potential lower than −0.7 MPa (Lu et al., 2006). Germination and radicle plus hypocotyl length in *Senna obtusifolia* (L.) Irwin and Barneby decreased with decreasing of solution osmotic potential, and no germination occurred at a solution osmotic potential of −0.75 MPa (Norsworthy and Oliveira, 2006). An osmotic potential between 0 and −0.2 MPa was found as optimum for seed germination of *S. sarrachoides*, where germination exceeded 90% (Zhou et al., 2005).

Water stress and drought may delay seeds germination in the field (Chauhan and Johnson, 2008c). Also, range of weed spread over habitats may be restricted to their seeds germination response to soil moisture conditions. Bouwmeester and Karssen (1993) found out desiccation may stimulate germination in *C. album* seeds interacting with nitrate in darkness. Thus, fluctuation in soil water content may stimulate seed germination in some species while it may restrict the germination of other species.

Seed response to water potential for germination depends on changes in dormancy level, as exhibited for temperature (see Figure 1). Huarte and Benech-Arnold, (2005) stated that the dormancy release may shifts the mean base water potentials needed for germination to more negative values, allowing the seed to germinate at a wider range of water potentials. Therefore, factors that govern changes in dormancy of seed populations like stratification and after-ripening, and those that terminate it like alternating temperatures and gibberellic acid, may operate through a reduction of mean base water potentials.

3.1.4. Soil Salinity

Soil salinity appears to have an adverse effect on seed germination. However, species are different in their response to salinity. For example, germination of *E. adenophorum* was inhibited at 300 mM NaCl and germination was greater than 65% at NaCl concentrations less than 100 mM (Lu et al., 2006). Chauhan et al., (2006d) reported that germination of *R. rugosum* seeds was relatively unaffected at a low level of salinity (80 mM NaCl), and some germination (11%) occurred even at 160 mM NaCl. At 160 mM NaCl, germination of *S. elaeagnifolium* was reduced to 5%, (Stanton et al., 2012). *Caperonia palustris* germination was greater than 52% at less than 40 mM NaCl and was lowest (27%) at 160 mM NaCl. Thus, *C. palustris* seed germination is fairly sensitive to NaCl concentration and seeds may germinate even at high soil salinity (Koger et al., 2004). *Commelina benghalensis* seeds were not tolerant to ≥10 mM NaCl for germination (Sabila et al., 2012). Seeds of *A. viridis* were more sensitive to increasing salt and water stress than *A. spinosus* for germination. *Amaranthus spinosus* germinated at a NaCl concentration of 100 mM (19%), whereas *A. viridis* seeds did not germinate at this concentration (Chauhan and Johnson, 2009b).

3.1.5. Nitrate

Nitrate is another factor promoting seed germination in many species. Potassium nitrate stimulated germination of (at 0.02 M) *S. orientale* (Chauhan et al., 2006c) and (at 0.005 M) *Malva parviflora* (Chauhan et al., 2006e). Goudey et al., (1988) demonstrated that nitrate can either stimulate or inhibit germination of dormant seeds of *S. arvensis*. Maximal germination frequencies were recorded in seed NO^{-}_{3} levels ranged from 0.3 to 4.4 nmol $seed^{-1}$. Germination was significantly lower in seeds containing more than 5 nmol NO^{-}_{3}. KNO_3 plus NH_4Cl were highly effective to stimulate the germination of *S. arvensis* especially when combined with irradiation and exposure to high temperature e.g. 20 °C (Goudey et al., 1978). However, Derkx and Karssen (1994) reported that nitrate or sensitivity to nitrate do not contribute to the regulation of dormancy and germination in *A. thaliana*. Potassium nitrate and ethephon can also promote seed germination in *A. fatua* especially in old seeds or seeds with lower dormancy. However, both potassium nitrate and ethephon at higher concentrations of 200 mM and 1 mM, respectively, caused reduction in seed viability (Adkins, 1993). Dissolved potassium nitrate (0.02 to 0.2% w/v) did not influence germination of *A. retroflexus* (Schonbeck and Egley, 1980).

3.1.6. Gibberellic Acid

Gibberellic acid (both external and internal in the seeds) affects seed germination. Germination of dormant seeds of *B. diandrus* (Kleemann and

Gill, 2012) and *H. murinum* ssp. *glaucum* populations increased with the addition of gibberellic acid (Fleet and Gill (2012). Derkx and Karssen (1994) observed that sensitivity to gibberellic acid to some extend contributes to seasonal dormancy pattern and germination in *A. thaliana*.

3.1.7. Gasses

Seed germination may affected by gaseous environment in the soil. Induction of seed dormancy seems to be mediated by lack of O_2 (Benvenuti and Macchia 1995, 1998) or by increase in CO_2 (Holm 1991). Germination of *Senna obtusifolia* was shown to be greater at 20% oxygen than at 2% oxygen (Norsworthy and Oliveira, 2006). Ethylene (1 to 100 ppmv) enhanced germination of *A. retroflexus* at 30 °C regardless of degrees of after-ripening, although the ethylene effect was most dramatic in nonafter-ripened seeds. Ethylene at 100 ppmv caused 40% germination in these seeds, compared to 1% for controls. However, carbon dioxide (0.001 to 4.5% v/v) did not influence germination (Schonbeck and Egley, 1980). In particular, germination may be inhibited in seeds positioned in compacted or very wet soil and in deeply buried seeds due to lack of enough oxygen required for the process of germination or high concentration of carbon dioxide due to biological activities in the soil.

3.1.8. PH

Soil PH is another factor affecting seed germination. There is also a range of pH that germination can take place. Germination in *P. oleracea* was greater than 70% over a pH range of 5 to 9, but declined to 57% at pH 4 (Chauhan and Johnson, 2008c). Germination of *H. hirta* at neutral pH was over 90%, and declined to 65% at pH 4 (acidic) and pH 10 (alkaline) levels (Chejara et al., 2008). The optimum pH range for germination of *S. sarrachoides* seeds was found to be between 6 and 8, although some seeds germinated at pH 4 and 9 (Zhou et al., 2005ab). Germination of *S. elaeagnifolium* occurred over a pH range of 4 to 10, but declined with increasing acidity (Stanton et al., 2012). Seeds of *Malva parviflora* germinated over a broad range of pH between 4 and 10, but germination declined at higher pH (Chauhan et al., 2006e). Seed germination of *R. rugosum* was greater than 76% over a pH range of 4 to 10 (Chauhan et al., 2006d). *Caperonia palustris* seed germination was greater than 42% over a pH range from 5 to 9. However, seed germination was declined to less than 32% at pH 4 and 10 (Koger et al., 2004). It is concluded that species are also different in their response to pH for germination.

4. Seedling Emergence

Knowledge of weed seedling emergence characteristics is of vital importance to consider in weed management strategies. The emergence characteristics of a weed species is assessed by emergence magnitude (number of emerged seedlings), time of emergence, and emergence pattern over time under experimental or field conditions. These all affected by soil depth, soil type, soil disturbance, crop residue, temperature and moisture condition of the soil or precipitation events and seed mass, (Dyer, 1995; Vleeshouwers and Kropff, 2000; Dieleman et al., 2000; Penny and Neal, 2003; Benvenuti et al., 2001; Benvenuti, 2003; Taab and Andersson, 2009c; Taab et al., 2011ab). Although these factors in fact may influence seed dormancy (e.g. temperature) and germination (e.g. temperature and water), they are explained here in seedling recruitment point of view. Seed production, viability and soil depth that seedlings of many important weeds species may potentially emerge are summarized in Table 2.

4.1. Soil Depth

In general, the number of emerged seedling is decreased with increasing of soil depth (Kremer and Lotz, 1998; Zhou et al., 2005; Chauhan and Johnson, 2008c). There is also a maximum soil depth from which seedlings of each weed species potentially may emerge (Table 2). Benvenuti et al. (2001) reported that burial depth has a great influence on mean emergence time. Each 2 cm increase in soil depth below the soil surface result in a significant increase in emergence time. This delay in emergence from deeper soil layers means that deeply buried seeds have a disadvantage in competition with crops. In addition, the depth from which germinated seeds can emerge determines which part of the soil seed pool may produce seedlings (Kremer and Lotz, 1998).

Seedling emergence in *P. oleracea* decreased sharply with increased seed burial depth. Seedling emergence was greatest (63%) for seed present on the soil surface, but decreased progressively as depth increased. No seedlings emerged from seed buried at a depth of 2 cm (Chauhan and Johnson, 2008c). Emergence fraction of *S. nigrum* biotypes was found to increase with temperature and decrease with increasing depth of placement in a sandy loam soil. Emergence fraction hardly changed from 1 to 2 cm depth of burial, but strongly decreased from 2 to 4 cm depth. No emergence was observed from a depth of 6 cm in their experiment. Emergence rate increased with increasing temperatures and decreasing depths (Kremer and Lotz, 1998). Benvenuti et al. (2001) reported that seedlings of *S. nigrum* emerge to a depth of 8 cm in a silt-loam soil, but emergence reduces below 50% by 5 cm soil depth. However, according to Keeley and Thullen (1983) seedlings of *S. nigrum* primarily emerge to a soil depth of 2.5 cm. *Solanum nigrum* seedlings were emerged from a soil depth of 10 cm (Taab, unpublished data). Zhou et al. (2005) reported maximum emergence of *S. sarrachoides* seedlings from planting depths of 2 cm or less. Seedling emergence decreased with increasing soil depth and no emergence occurred at soil depths of more than 8 cm.

Seedlings of *Gutierrezia sarothrae*, *G. microcephala*, *Isocoma coronopifolia*, and *Ericameria austrotexana* emerged readily if placed on the soil surface or partially pressed into the soil. Emergence was reduced by covering seed with soil to a depth of less than 1 mm. Few seedlings emerged from a depth of 1 cm, and none emerged from a depth of 2 cm. Rate of emergence was also decreased with increasing depth of placement. Soil texture had little effect on emergence of these species (Mayeux Jr, 1983). Seedling emergence of *Mimosa invisa* was 80 to 94% at depths of 0 to 2 cm but decreased progressively at deeper depths, and no seedlings emerged from seed buried at 10 cm (Chauhan and Johnson, 2008a). Maximum seedling emergence of *Trianthema portulacastrum* L. was resulted when seeds were planted at 1 cm and gradually declined with the increased seeding depth. Seeds lying on the surface gave very poor seedling emergence (Balyan and Bhan, 1986). Seedling emergence of *Verbesina encelioides* was highest from surface-sown seeds and least in those sown at 2.5 cm soil depth, beyond which no emergence occurred (Kaul and Mangal, 1987).

Seedling emergence of *A. spinosus* and *A. viridis* were 56 and 68%, respectively, for seeds placed on the soil surface. Only 7% of *A. spinosus* seedlings emerged from a soil depth of 0.5 cm, whereas no emergence was observed from 4 cm or deeper. For *A. viridis*, 6% emergence was observed at 4 cm depth and no seedling emerged from 6 cm depth (Chauhan and Johnson, 2009b). Moreover, maximum seedling emergence of *A. retroflexus* (33±3.2%) was observed for seeds placed at soil surface. Thereafter, seedling emergence was decreased with depth and reached its minimum at the depth of 6(2.3±1%) cm, followed by surprising increased emergence at the depths of 8(3.3±1.2%) and 10(12.3±2.2%) cm (Taab et al, 2011a). A similar pattern was also observed in the second year of the study (Taab, unpublished data).

Table 2. Seed production, viability and seedling emergence from soil depths in some weed species

Species	Seed production/ plant	Seed viability (year)	Emergence depth (cm)	Reference
Amaranthus albus	6,000,000	40	6-8	Sokolova, 2008
A. blitoides	700,000 -	40 -	8 10	Sokolova, 2008 Taab et al., 2011a
A. retroflexus	5000	40	- 10	Sokolova, 2008 Taab et al., 2011b
Ambrosia artemisiifolia	1000 – 100,000	40	8	Nadtochii 2008
Apera spica-venti	-	6-7	<2	Luneva, 2008
Artemisia vulgaris	15,000		<3	Nadtochii 2008
Avena fatua	500	15	-	Luneva, 2008
Brassica juncea	20,000	5	6	Larina, 2008
Capsella bursa-pastoris	70,000	11	2-3	Larina, 2008
Cardaria draba	3000-5000	2	2-3	Larina, 2008
Centaurea cyanus	7000	3	1-6	Nadtochii 2008
Chenopodium album	600,000	10	- 10	Larina, 2008 Taab et al., 2011Ch2
Convolvulus arvensis	9800 -	50-55	10 -	Nadtochii 2008 Brown and Porter, 1942
Cuscuta campestris	100,000	3-10	-	Doronina, 2008
Echinochloa crusgalli	200-1000	7-10	12	Luneva, 2008
Galium aparine	1200	5	8-9	Doronina, 2008
Helianthus lenticularis	19,000	2-3	5-20	Nadtochii 2008
Plantago major	61,000 - 320,000	7	3	Nadtochii 2008
Portulaca oleracea	1,000,000 -	30 -	1.5 6	Nadtochii 2008 Benvenuti et al., 2001
Raphanus raphanistrum	2500 - -	10 or 15 -	3-4 - 8	Larina, 2008 Murrumbidgee CMA 2012 Young and Cousens, 1999
Reseda lutea	400,000	5	4	Nadtochii 2008
Rumex acetosella	10,000	5	8-10	Doronina, 2008
R. crispus	7000		5-7	Doronina, 2008
Sinapis arvensis	20,000	10	5-6	Larina, 2008

Table 2. (Continued)

Species	Seed production/ plant	Seed viability (year)	Emergence depth (cm)	Reference
Solanum nigrum	600 - 168,000	18-39	-	Keeley and Thullen, 1983; Ogg and Rogers, 1989; Holm et al. 1991; Defelice 2003;
	-	-	10	Taab, unpoblished
Stellaria media	10,000-20,000	2-5	1-2	Sokolova, 2008
Thlaspi arvense	50,000	10	1	Larina, 2008
Viola tricolor	3020	6	4-5	Nadtochii 2008

Maximum seedling emergence of *A. blitoides* (46±6.6%) was also observed for seeds at soil surface and emergence decreased with increasing depth and reached its minimum at the depths of 8(13±3.2%) and 10(14.8±3.4%) cm (Taab et al, 2011a). In the next year of the study, the emergence was high (42.8%) for seeds placed at the soil surface, decreased at the depth of 2 and 4 cm followed by subsequent increase at the depth of 6, 8, and maximum emergence (55.5±14%) at the depth of 10 cm (Taab, unpublished data).

Seedling emergence in some species is inhibited for seeds placed at soil surface. For example, seedling emergence of *Veronica persica*, *V. arvensis*, *S. media*, *C. album*, and *P. aviculare* was inhibited in seeds placed at the soil surface, it increased at the shallower depths followed by subsequent decline at the deeper depths (Grundy *et al.*, (2003a). A similar response was also observed for seedling emergence of *Carthamus oxyacantha, Vaccaria pyramidata* (see Taab et al., chapter 2*)* and *C. album* (Bliss and Smith, 1985). However, a different seedling emergence response was observed in *C. album* seeds (see Taab et al., chapter 3). Therefore, seedling emergence response to soil depth is mainly a species-specific response, but there might be other factors affecting this response e.g. differences in population adaptation characteristics.

Thus, the fact is species are different in their seedling emergence response to soil depth. Moreover, there is usually a combination of several species that could be found in a field. These make weed seed pool management a difficult task to do. However, the more we know about the functional behavior of the species, the better we can take action to manage them in agriculture.

4.2. Crop Residue

Crop residue at the soil surface may affect weed seedling emergence substantially (Dyer, 1995; Haramoto and Gallandt, 2005; Chauhan and Johnson, 2009a) depending on weed species. Weed emergence may decrease by increasing of mulch depth like planting depth (Penny and Neal, 2003). Higher stubble levels increased the emergence of larger seeded species including *A. fatua* and *Echinochloa colona* (L.) Link, but decreased the emergence of *Sonchus oleraceus* L. (Widderick et al., 2010). Dyer (1995) stated that higher levels of crop residue increasingly reduce and delay weed seedling emergences, most likely by decreasing soil temperatures and preventing light penetration. Residues from *Brassica* cover crops reduced emergence of weeds by 23 to 34% compared with fallow, and weed emergence was also delayed by approximately 2 days (Haramoto and Gallandt, 2005). Glucosinolates from oilseed meals that hydrolyze into biologically active compounds was shown to potentially suppress and control *Sorghum halepense* Pers. and *A. retroflexus* (Rothlisberger et al., 2012).

4.3. Seed Mass

The amount of food reserve in a seed and therefore seed weight determine the maximum depth at which seedlings may emerge (Kremer and Lotz, 1998). Benvenuti et al. (2001) found a significant correlation between seed specific weight and depth-mediated 50% emergence inhibition among 20 weed species. Their results indicated that seed weight can represent a valid tool for identifying the emergence profile of the active soil seed bank i.e. likelihood of emergence and consequent competition with crop. Germinability and the rate of germination are related to seed mass in some plant species. There is, sometimes, a pronounced fluctuation in seed unit weight presents even within the same species (Milberg et al. 1996). This variation could affect emergence characteristics of a species (Benvenuti et al., 2001).

4.4. Seed Populations

Kremer and Lotz (1998) reported significant differences between triazine resistant and susceptible *S. nigrum* biotypes in emergence fraction. Resistant seeds showed a higher emergence fraction than susceptible seeds while emergence rate of resistant seeds was slightly lower. Differences were also found in total seedling emergences among populations of *A. ludoviciana* and *A. fatua* (Taab and Striebig, 2005).

A non-dormant population of *H. murinum* ssp. *glaucum*, collected from a long-term pasture (non-crop habitat), showed high seedling emergence (>90%) during autumn, which was well before planting of the winter crop (lentil). In contrast, populations sampled from cropping fields showed very little seedling establishment (<10%) before crop planting. It is difficult to control this species in cereals because there are no selective herbicides available for the control of weedy barley species (Fleet and Gill (2012).

Differences in seed dormancy among *B. diandrus* populations were expressed in seedling emergence pattern in the field. The non-dormant populations collected from fencelines, showed high seedling establishment (>80%) during autumn, which coincided with the planting time of winter crops in southern Australia. In contrast, five populations from cropping fields showed much lower seedling establishment (3 to 17%) before the time of crop planting. Delayed seedling establishment in populations from cropping fields could lead to less effective pre-seeding weed control and higher weed infestations in field crops. Results of this study also showed that the seedbank of these highly dormant *B. diandrus* populations can readily persist from one year to the next (Kleemann and Gill (2012).

4.5. Soil Type

Seedling emergence is influenced by soil texture. For example, total weed seedling emergence in sandy loam was higher than clay loam soil (Gulshan and Dasti, 2012). Emergence percentage was also lower in clay than in clay loam or sandy clay loam soil (Mutz and Scifres, 1975). Soil physical properties have a strong effect on seed ecology and consequently on seed bank dynamics (Benvenuti, 2003). Soil physical characteristics determine the soil penetration resistance for the seedlings to reach the soil surface (Vleeshouwers, 1997a) and consequently may affect the emergence timing e.g. cause delayed emergence in heavy soils.

4.6. Moisture

Variation and patchiness in available soil moisture and soil fertility that would influence weed seed germination, emergence, and seedling growth and reproduction may cause weed patchiness (Dieleman et al., 2000). There is usually a seedling emergence flush after rainfall events more pronounced especially in rainfed crops (personal observations). Widderick et al., (2010) showed that irrigation, with timing of irrigation being important, increases the emergence of *S. oleraceus, E. colona*, *E. crus-galli*, *Urochloa panicoides???* and *A. fatua*. Therefore, changes in the soil moisture (explained as water potential) is considered as an important factor to develop predictive tools for predicting seedling emergence timing in the field, especially where the soil moisture is a significant determinant of seedling emergences.

4.7. Temperature and Emergence Pattern

There is a time period during which seedlings of a weed species emerge in the field, interacting with the environment (Taab, 2009; Taab et al., 2011ab) depending on season and adaptation characteristics of the species. Seedling emergences of *S. nigrum* usually take place from mid February until mid May with a peak in late March to early April in Iran (Taab, unpublished data). Taab and Andersson (2009c) showed seedling emergence in populations of *S. nigrum* and *S. physalifolium* from early May until early July with declined emergence after this period in Sweden. In New Zealand, Popay *et al*., (1995) reported emergence of *S. nigrum* in late spring and summer with some emergence continuing into autumn. According to a study by Roberts and Lockett (1978) in the UK, seedling of *S. nigrum* begin to emerge in early May, with the main peak in late May, continued during June and July. Only a few seedlings appeared in August and none at all after mid-September. A study by Ogg and Dawson (1984) showed that *S. nigrum* seedlings generally began to emerge during the first 2 weeks of April and emergence generally peaked in mid-April to mid-May and continued until September in the USA. Keeley and Thullen (1983) reported that *S. nigrum* seedlings began to emerge in March, in California, when soil temperature at 5 cm depth reaches 17°C.

Emergence timing of weed species seems to be mainly regulated by the soil temperature (Taab, 2009; Taab et al., 2011ab). The emergence of seedling flushes of *S. nigrum* and *S. physalifolium* was attributed to a daily mean soil temperature above 13 to 17°C with a ca 10°C fluctuation (Taab and

Andersson, 2009c). A similar result was obtained for *S. nigrum* in Iran (Taab, unpublished data). A requirement of a fluctuating temperature above 15 to 20°C for seedling emergence of *S. nigrum* has also been reported (Edmonds and Chweya, 1997; Roberts and Lockett, 1978; Keeley and Thullen, 1983; Ogg and Dawson, 1984). The emergence timing of *S. nigrum* and *S. physalifolium* was closely resembles the pattern of seasonal dormancy cycle in the field when dormancy induction after early July prevents further seed germination (Taab and Andersson, 2009a).

Taab et al., (2011a) reported that seedling emergence of *A. retroflexus* started sporadically in February when mean daily soil temperature increased above 5°C. The seedling emergence reached its peak in the end of March and early April and tailed off in early May. For *A. blitoides,* seedling emergence was also started sporadically in February, with its peak in the beginning of April and in the end of March and tailed off in mid May (Taab et al., 2011b). Seedling emergences in both species was coincided with the increase of soil temperature above 10°C with ca 5 to 25°C fluctuation in the preceding weeks. Moreover, seedling emergence tailing off was also coincided with the increasing of daily mean soil temperature above 20°C indicating induction of dormancy (Taab et al., 2011ab). A soil temperature of 15 to 20 C was found best to achieved maximum emergence of *Sisymbrium irio* (Ray et al., 2005).

Vleeshouwers (1997b) defined the percentage of seedling emerging, their mean time of emergence and the spread around this mean time of emergence as the three most important parameters describing the time course of seedling emergence. Seedlings of some weed species may emerge with a main flush over a short period of time i.e. *C. oxyacantha* and *V. pyramidata* (see Taab et al., chapter 2). However, many weed species showed bimodal pattern of seedling emergence in the field (Ogg and Dawson 1984; Myers et al. 2005; Stoller and Wax 1973). Taab and Andersson (2009c) reported an extended bi- or three-modal recruitment pattern in addition to sporadic seedling emergence in *S. nigrum* and *S. physalifolium.* They attributed this extended emergence pattern to variation in dormancy state among seed batches reflected in germination timing. Seeds with different level of dormancy have different requirement for dormancy release (Taab and Andersson, 2009b). Temperature, soil penetration resistance and burial depth was shown to affect the percentage and the mean time of emergence in *P. persicaria*, *C. album* and *S. arvensis.* Fraction of emergence decreased with increasing burial depth and increasing penetration resistance. The rate of emergence increased with increasing temperature, decreasing burial depth and decreasing soil penetration resistance in these species (Vleeshouwers, 1997a).

4.8. Soil Disturbance

Substantial weed seedlings may emerge as a result of soil cultivation. Soil cultivation triggers only the first flush of seedlings following soil cultivation, and later flushes and flushes in undisturbed soils are triggered by other factors like desiccation and remoistening of the seeds (Vleeshouwers and Kropff, 2000) or fluctuating temperature (Taab, 2009) in the upper soil layers.

Widderick et al., (2010) reported up to 70% reduction in the *A. fatua* seed bank in the top 5 cm of soil either with or without soil cultivation. However, cultivation increased the number of *A. fatua* seeds persisting below 5 cm. Cultivation was also appeared to increase the persistence of *S. oleraceus, E. colona*, *E. crus-galli*, and *U. panicoides*. A study has shown that emergence of small-seeds broadleaf weed species is often lower when tillage is conducted during darkness than tillage during the light. Moreover, annual grass and large-seeds broadleaf species showed little response to light environment during tillage (Buhler, 1997).

Low-soil-disturbance tillage systems may leave more seeds on the soil surface, whereas the high-soil-disturbance systems bury most of the seeds. The seedling recruitment of *L. rigidum* was shown to be lower under the low-soil-disturbance tillage systems than under the high-soil-disturbance tillage systems. The seedling emergence under minimum tillage was greater than under no-till (Chauhan et al., (2006b). In addition, Rahman et al., (2000), comparing different soil disturbance measures, concluded that ploughing cause a significant shift of seeds to the deeper soil profile than power harrow and rotary hoe. Ploughing could result in significantly less weeds than rotary hoeing or harrowing. They pointed out that summer grasses are likely to be a bigger problem in no-tillage systems than broadleaf weeds because they will emerge earlier and start competing with the crop (Rahman et al., 2000).

4.9. Species Interaction

Seedling emergence and initial survival of *Koelaria cristata*, *Bouteloua gracilis*, and *Sporobolus cryptandrus* were decreased by the presence of *Acroptilon repens* roots. In contrast, survival and growth of *Agropyon smithii* were unaffected by the presence of *A. repens* (Grant et al., 2003). Therefore, weed species may affect germinability and emergence of each other and there might be differences among species in their response to allelo-chemicals.

5. Ecological Implication

Seed dormancy is an important adaptation characteristics functioning to regulate germination to time seedling emergence to appropriate condition for seedling establishment and subsequent reproduction. For example in case of summer annual species, seed dormancy is alleviated during autumn, winter and early spring so that seeds would be able to germinate in spring or early summer. Thereafter, emerged seedlings would have enough time for growth and completing their life cycle before autumn. Moreover, induction of secondary dormancy during summer prevent late emergence otherwise the plants would die in autumn due to low temperature before reproduction. Dormancy is reduced in the seasons preceding suitable conditions for weed emergence in species that show seasonal changes in their seed dormancy (Batlla and Benech-Arnold, 2007). Dormancy in *Ambrosia trifida* L. seeds, may limit potential for fatal germination during fall. Moreover, late-season emergence in *A. trifida* involves high levels of embryo dormancy that prevent germination at low temperatures in spring (Schutte et al., 2012).

The formation of long-lived, heterogeneous seed pools in the soil is the inevitable consequence of the dormant seed rain (Dekker, 2003). Seed dormancy also spreads the germination over time e.g. more than one season or several years (Silvertown and Charlesworth, 2006). Therefore, if a natural catastrophic event happens in one season, portions of the dormant seeds remain in the soil to guarantee the future reproduction of the species. The production of heterogeneous dormant seeds by an individual plant is a strategy to ensure the survival of next generations (Matilla et al., 2005). Mother plant produce seeds with different levels of dormancy predicting the future necessity of distribution in emergence timing of the species to reduce risks associated with growth and survival and to maximize the chance of plant establishment and reproduction. Intense competition avoidance is also a consequence of variation in dormancy, so that seeds with different level of dormancy need different conditions and time to overcome dormancy to germinate.

6. Agricultural Implications

Weed control operation in early season is of great importance because weed seedlings are not large enough yet to affect crops by taking up the common resources. To do so, information on weed seedling emergence timing is needed. Moreover, to develop efficient integrated weed management strategies, knowledge of emergence characteristics is required (Kremer and Lotz, 1998). Dormancy is a seed characteristic that regulates germinability and emergence timing of many weed species. Number of established plants is also determined by portion of the seeds released from dormancy (Benech-Arnold et al., 2000) in the soil seed pool. Thus, prediction of dormancy degree of buried weed seed population could be a good indication of possible number of emerged seedling after cultivation (Vleeshouwers and Kropff, 2000). The weed seedlings that emerge before or simultaneous to the crop may have a competitive advantage. Therefore, a high priority need to be given to the control of early weed flushes in weed management systems. Reduction of dormancy causes gradual increase in seeds sensitivity to environmental signals e.g. light and temperature, for germination. The timing of seedling emergence depends on the rate of germination and pre-emergence growth (Vleeshouwers and Kropff, 2000). Dekker (2003) studying the *Setaria* species group, pointed out that precise and variable timing of seedling emergence ensures *Setaria* species dominant place in disturbed and managed communities during the growth and reproductive phases that follow.

Manipulating the crop sowing date in relation to weed seedling emergence timing can be taken into consideration as a good measure to control weeds. This enables the farmers to remove portion of the emerged seedling early in the season using seed bed preparation and sowing operation. Soil tillage during light increases emergence of weeds especially in small-seeded broadleaf species. However, soil disturbance (Popay et al., 1994) in particular during light (Buhler, 1997) is known as a tool to stimulate seedling emergence and consequently reduce the weed seed pool in the field.

Different soil disturbance measures can be used to affect seed distribution in the soil profile. For instance, shallow tollage leave more seeds in the upper soil layers. In contrast, deep soil disturbances may shift more seeds to the deeper soil layers (Rahman et al., 2000; Chauhan et al., (2006b) resulting in disadvantageous delayed seedling emergence or ensuring long persistent of seeds in the soil seed pool by preventing germination. There is also a high risk of predation for seeds at the soil surface that could be an option to reduce seed input to the soil seed pool.

Crop residues at the soil surface can reduce and delay weed seedling emergence (Dyer, 1995; Haramoto and Gallandt, 2005). In addition, compounds from the residues may suppress the weeds (Rothlisberger et al., 2012). Thus, crop residue at the soil surface can be taken into account as a weed control option to reduce weed seedling emergences and suppress the weeds.

It should be noted that in crops growing with wide row spacing and relatively slow canopy closure the crop may not be able to affect weed seedling emergence and growth (Leblanc et al., 2002). This condition is likely to create an opportunity space for weed seedling emergence and establishment, which has to be taken into account in crop sequences and weed management decisions.

CONCLUSION

Seed dormancy (especially physiological dormancy) has a profound influence on regulating germinability and subsequent emergence timing of weed species. All environmental signals during mother plant growth may contribute in formation and degree of dormancy. The weakly dormant seeds need a shorter period of time to release dormancy and the strongly dormant seeds need longer period of time to release dormancy. Heterogeneous dormant seed rain by parent plants is likely to be a key adaptation characteristic suggesting prediction of uncertainty in the environment for successful establishment and survival of the species. This characteristic enables the species to distribute germination and emergence timing over time. Temperature is known as the most important factor affecting dormancy and germination. There is also a possible interaction between temperature and other factor e.g. light, for germination to take place. In addition, dormancy affect the condition required for germination. For example, dormancy reduction causes reduction in requirement for alternating temperature or light for germination. There is also a range of environmental conditions, (e.g. temperature, water potential, salinity, pH etc.) within which seeds of a species may germinate. These ranges could be different among the species and thus species-specific. Recently, there are observed uncertainties and changes in some environmental factors like temperature and precipitations due to the climate changes, more pronounced in some parts of the world. This demands to develop precise predictive tools to predict weed seedling emergence timing and behavior in the ecosystems. In addition, there are progressive limitations on using herbicides in crop production systems due to human health and environmental hazards, which demand adapting integrated approach to tackle weeds. In particular, weed control operation in the early stage of growth is of great importance when weed seedlings would not be able to affect the crop significantly. Therefore, information on weed seed dormancy, germination and seedling emergence as a key factor need to be considered in integrated weed management strategies.

REFERENCES

Adkins S.W. (1993) Latitudinal differentiation in heading photoperiod sensitivity and seed dormancy of *Echinochloa oryzicola*, an obligate weed in flooded rice. Proceeding of the 10th Australian Weeds Conference/14th Asian-Pacific Weed Science Society Conference, 6-10 September 1993, Brisbane, Queensland, Australia, VL (1) pp 414-418.

Andersson L. and Åkerblom-Espeby L (2009) Variation in seed dormancy and light sensitivity in *Alopecurus myosuroides* and *Apera spica-venti. Weed Research* 49: 261-270.

Allen P.S. and Meyer S.E. (1998) Ecological aspects of seed dormancy loss. *Seed Science Research* 8: 183-192.

Andersson L. and Milberg P. (1998) Variation in seed dormancy among mother plants, populations and years of seed collection. *Seed Science Research* 8: 29–38.

Armstrong L.J., Adkins S.W. (1998) Variation in photoperiod response of different isogenic lines of wild oats (*Avena fatua*). *Weed Science* 46(1): 39-47.

Balyan R.S. and Bhan V.M. (1986) Germination of Horse Purslane (*Trianthema portulacastrum*) in Relation to Temperature, Storage Conditions, and Seeding Depths. *Weed Science,* 34(4): 513-515.

Baskin C.C. and Baskin J.M. (1998) *Seeds: Ecology, Biogeography, and Evolution of Dormancy and Germination.* 1st ed. San Diego, CA: Academic Press. pp. 27–124, 185–200.

Baskin J.M. and Baskin C.C. (1981) Seasonal changes in the germination responses of buried *Lamium amplexicaule* seeds. Weed Research 21: 299–306.

Baskin J.M. and Baskin C.C. (2004) A classification system for seed dormancy. *Seed Science Research* 14: 1–16.

Batlla D. and Benech-Arnold R.L. (2003) A quantitative analysis of dormancy loss dynamics in *Polygonum aviculare* L. seeds: development of a thermal time model based on changes in seed population thermal parameters. *Seed Science Research* 13: 55-68.

Batlla D. and Benech-Arnold R.L. (2006). The role of fluctuations in soil water content on the regulation of dormancy changes in buried seeds of *Polygonum aviculare* L. *Seed Science Research* 16: 47-59.

Batlla D. and Benech-Arnold R.L. (2007). Predicting changes in dormancy level in weed seed soil banks: implications for weed management. *Crop Protection* 26: 189-197.

Benech-Arnold R.L., Sanchez R.A., Forcella F., Kruk B.C. and Ghersa C.M. (2000) Environmental control of dormancy in weed seed banks in soil. *Field Crops Research* 67: 105–122.

Benvenuti S., Falorni C., Simonelli G. and Macchia M. (2001) Weed seedbank evaluation and relative emergence dynamics in three perenial medicinal crops of organic farming systems. *Italian Journal of Agronomy* 5:29-37.

Benvenuti S. and Macchia M. (1995) Hypoxia effect on buried weed seed germination. *Weed Research* 35: 343-351.

Benvenuti S. and Macchia M. (1998) Phytochrome mediated germination control of *Datura stramonium* L. seeds. *Weed Research* 38: 199-205.

Benvenuti S., M. Macchia, Miele S. (2001) Light, temperature and burial depth effects on *Rumex obtusifolius* seed germination. *Weed Research* 41:177–186.

Bithell S.L., McKenzie B.A., Bourdot G.W., Hill G.D. and Wratten S.D. (2002) Germination requirements of laboratory stored seeds of black nightshade and hairy nightshade. *New Zealand Plant Protection* 55: 222–227.

Bliss D. and Smith H. (1985) Penetration of light into soil and its role in the control of seed germination. *Plant, Cell and Environment* 8: 475–483.

Bouwmeester H.J. (1990) The effect of environmental conditions on the seasonal dormancy pattern and germination of weed seeds. PhD thesis, Wageningen Agricultural University, The Netherlands.

Bouwmeester H.J. and Karssen C.M. (1992) The dual role of temperature in the regulation of the seasonal changes in dormancy and germination of seeds of *Polygonum persicaria* L, *Oecologia* 90: 88–94.

Bouwmeester H.J. and Karssen C.M. (1993) Seasonal periodicity in germination of seeds of *Chenopodium album* L. *Annals of Botany* 72: 463–473.

Brainard D.C., Bellinder R.R. and DiTommaso A. (2005) Effects of canopy shade on the morphology, phenology, and seed characteristics of Powell amaranth (*Amaranthus powellii*). *Weed Science* 53: 175–186.

Brainard D.C., DiTommaso A. and Mohler C.L. (2006) Intraspecific variation in germination response to ammonium nitrate of Powell amaranth

(*Amaranthus powellii*) seeds originating from organic vs. conventional vegetable farms. *Weed Science* 54: 435–442.

Brainard D.C., DiTommaso A. and Mohler C.A. (2007) Intraspecific variation in seed characteristics of Powell amaranth (*Amaranthus powellii*) from habitats with contrasting crop rotation histories. *Weed Science* 55: 218–226.

Brändel M. (2005) The effect of stratification temperatures on the level of dormancy in primary and secondary dormant seeds of two *Carex* species. *Plant Ecology* 178: 163–169.

Brändel M. and Jensen K. (2005) Effect of temperature on dormancy and germination of *Eupatorium cannabinum* L. achenes. *Seed Science Research* 15: 143–151.

Brown E.O. and Porter R.H. (1942) The viability and germination of seeds of *Convolvulus arvensis* L. and other perennial weeds. Agricultural Exp. Station, Iowa State College, *Research Bulletin*, No. 294.

Buhler D.D. (1997) Effects of tillage and light environment on emergence of 13 annual weeds. *Weed Technology* 11: 496–501.

Casal J.J. and Sanchez R.A. (1998) Phytochromes and seed germination. *Seed Science Research* 8: 317–329.

Chauhan B. S., Gill G., and Preston C. (2006a). Factors affecting seed germination of threehorn bedstraw (*Galium tricornutum*) in Australia. *Weed Science* 54: 471-477.

Chauhan B.S., Gill G. and Preston C. (2006b) Influence of tillage systems on vertical distribution, seedling recruitment and persistence of Rigid Ryegrass (*Lolium rigidum*) Seed Bank. *Weed Science,* 54(4): 669-676.

Chauhan B.S., Gill G. and Preston C. (2006c) Influence of environmental factors on seed germination and seedling emergence of Oriental mustard (*Sisymbrium orientale*). *Weed Science* 54(6): 1025-1031.

Chauhan B.S., Gill G., and Preston C. (2006d) Factors affecting turnipweed (*Rapistrum rugosum*) seed germination in southern Australia. *Weed Science* 54 (6): 1032-1036.

Chauhan B.S., Gill G. and Preston C. (2006e) Factors affecting seed germination of little mallow (*Malva parviflora*) in southern Australia. *Weed Science* 54(6): 1045-1050.

Chauhan B.S. and Johnson D.E. (2008a) Seed Germination and Seedling Emergence of Giant Sensitiveplant (*Mimosa invisa*). *Weed Science* 56(2): 244-248.

Chauhan B.S. and Johnson D.E. (2008b). Influence of environmental factors on seed germination and seedling emergence of eclipta (*Eclipta prostrata*) in a tropical environment. *Weed Science* 56: 383-388.

Chauhan B. S. and Johnson D. E. (2008c) Germination biology of *Portulaca oleracea* L. 16th Australian Weeds Conference, Cairns, Queensland, Australia, 18-22 May 2008 183- 185

Chauhan B.S. and Johnson D.E. (2009a) Seed germination ecology of *Portulaca oleracea* L.: an important weed of rice and upland crops. *Annals of Applied Biology* 155: 61–69.

Chauhan B.S. and Johnson D.E. (2009b) Germination ecology of spiny (*Amaranthus spinosus*) and slender amaranth (*A. viridis*): Troublesome weeds of direct-seeded rice. *Weed Science* 57(4): 379-385

Cheam A.H. and Code G.R. (1998) *The Biology of Australian Weeds*, vol. 2, ed. F.D. Panetta, R.H. Groves and R.C.H. Shepherd, R.G. and F.J. Richardson, Melbourne.

Chejara V. K., Kristiansen P., Whalley R. D. B., Sindel B. M., and Nadolny C. (2008) Factors affecting germination of coolatai grass (*Hyparrhenia hirta*). *Weed Science* 56(4) 543-548.

Chen P.H. and Kuo W.H.J. (1999) Seasonal changes in the germination of buried seeds of *Monochoria vaginalis*. *Weed Research* 39: 107–115.

Cirujeda A., Recasens J. and Taberner A. (2006) Dormancy cycle and viability of buried seeds of *Papaver rhoeas*. *Weed Research* 46: 327-334.

Clua A., Fernandez G., Ferro L. and Dietrich M. (2006) Drought stress conditions during seed development of narrowleaf birdsfoot trefoil (*Lotus glaber*) influences seed production and subsequent dormancy and germination. *Lotus Newsletter* 36 (2): 58-63.

Cone J.W. and Spruit C.J.P. (1983) Imbibition conditions and seed dormancy of *Arabidopsis thaliana*. *Physiologia Plantarum* 59: 416-420.

Contreras S.A. (2007) Effects of maternal plant environment on (*Lactuca sativa* L.) seed dormancy, germinability, and storability. PhD dissertation, The Ohio State University, USA.

Dasti A.A., Fatima K., and Malik S.A. (2001) Storage time on seed dormancy and germination in Eti mutant of *Arabidopsis thaliana* (L) Heynh. *Journal of Research (Science),* 12(1): 34-42.

Defelice M.S. (2003) The black nightshade, *Solanum nigrum* L. et al.-Poison, poultice, and Pie. *Weed Technology* 17: 421-427.

Dekker J. (2003) The foxtail (*Setaria*) species-group. *Weed Science* 51(5): 641-656.

Del Monte J.P. and Tarquis A.M. (1997) The role of temperature in the seed germination of two species of the *Solanum nigrum* complex. *Journal of Experimental Botany* 48: 2087–2093.

Derkx M.P.M. and Karssen C.M. (1994) Are seasonal dormancy patterns in *Arabidopsis thaliana* regulated by changes in seed sensitivity to light, nitrate and gibberellin? *Annals of Botany* 73, 129–136.

Dieleman J.A., Mortensen D.A., Buhler D.D. and Ferguson R.B. (2000) Identifying associations among site properties and weed species abundance. II. Hypothesis Generation. *Weed Science* 48(5):576-587.

Donohue K., Heschel M.S., Chiang G.C.K, Butler C.M and Barua D. (2007) Phytochrome mediates germination responses to multiple seasonal cues. *Plant Cell Environment* 30(2): 202-212.

Dorado J., Fernández-Quintanilla C. and Grundy A.C. (2009) Germination patterns in naturally chilled and nonchilled seeds of Fierce Thornapple (*Datura Ferox*) and velvetleaf (*Abutilon Theophrasti*). Weed Science 57(2): 155-162.

Doronina A.Yu. (2008) *Cuscuta campestris*. In A.N. Afonin, S.L. Greene, N.I. Dzyubenko, A.N. Frolov (eds.). 2008. Interactive Agricultural Ecological Atlas of Russia and Neighboring Countries. Economic Plants and their Diseases, Pests and Weeds [Online]. Available at: http://www. agroatlas.ru/en/content/weeds/Cuscuta_campestris/

Doronina A.Yu. (2008) *Fallopia convolvulus*. In A.N. Afonin, S.L. Greene, N.I. Dzyubenko, A.N. Frolov (eds.). 2008. Interactive Agricultural Ecological Atlas of Russia and Neighboring Countries. Economic Plants and their Diseases, Pests and Weeds [Online]. Available at: http://www. agroatlas.ru/en/content/weeds/Fallopia_convolvulus/

Doronina A.Yu.. (2008) *Galium aparine*. In A.N. Afonin, S.L. Greene, N.I. Dzyubenko, A.N. Frolov (eds.). 2008. Interactive Agricultural Ecological Atlas of Russia and Neighboring Countries. Economic Plants and their Diseases, Pests and Weeds [Online]. Available at: http://www. agroatlas.ru/ en/content/weeds/Galium_aparine/

Doronina A.Yu. (2008) *Rumex acetosella*. In A.N. Afonin, S.L. Greene, N.I. Dzyubenko, A.N. Frolov (eds.). 2008. Interactive Agricultural Ecological Atlas of Russia and Neighboring Countries. Economic Plants and their Diseases, Pests and Weeds [Online]. Available at: http://www. agroatlas.ru/en/content/weeds/Rumex_acetosella/

Doronina A.Yu. (2008) *Rumex crispus*. In A.N. Afonin, S.L. Greene, N.I. Dzyubenko, A.N. Frolov (eds.). 2008. Interactive Agricultural Ecological Atlas of Russia and Neighboring Countries. Economic Plants and their

Diseases, Pests and Weeds [Online]. Available at: http://www.agroatlas.ru/ en/content/ weeds/Rumex_crispus/

Dyer W.E. (1995) Exploiting weed seed dormancy and germination requirements through agronomic practices. *Weed Science* 43:498-503.

Edmonds J.M. and Chweya J.A. (1997) *Black nightshades (Solanum nigrum L.) and related species. Promoting the conservation and use of underutilized and neglected crops.* 15. Institute of Plant Genetics and Crop Plant Research, Gatersleben/International Plant Genetic Resources Institute, Rome, Italy.

Esashi, Y., Kuraishi, R., Tanaka, N. and Satho, S. (1983) Transition from primary dormancy to secondary dormancy in cocklebur seeds. *Plant, cell andandandandand environment* 6, 493–499.

Fandrich L. and Mallory-Smith C. A. (2006) Factors affecting germination of jointed goatgrass (*Aegilops cylindrica*) seed. Weed Science 54(4): 677-684.

Fawcett R.S. and Slife F.W. (1978) Effects of field applications of nitrate on weed seed germination and dormancy. *Weed Science* 26: 594–596.

Fenner M. (1991) The effects of the parent environment on seed germinability. *Seed Science Research* 1: 75–84.

Figueroa R., Herms D.A., Cardina J. and Doohan D. (2010) Maternal environment effects on common groundsel (*Senecio vulgaris*) seed dormancy. *Weed Science* 58:160–166.

Finch-Savage W.E. and Leubner-Metzger G. (2006) Tansley review: Seed dormancy and the control of germination. *New Phytologist* 171: 501-523.

Finkelstein R., Reeves W., Ariizumi T. and Steber C. (2008) Molecular aspects of seed dormancy. *Annual Review of Plant Biology* 59: 387-415.

Fleet B. and Gill G. (2012) Seed dormancy and seedling recruitment in smooth barley (*Hordeum murinum* ssp. *glaucum*) populations in southern Australia. Weed Science 60(3): 394-400.

Foley M.E. (2001) Seed dormancy: an update on terminology, physiological genetics, and quantitative trait loci regulating germinability. *Weed Science* 49:305–317.

Foley M.E. (2004) Leafy spurge (*Euphorbia esula*) seed dormancy. *Weed Science* 52(1): 74-77.

Foley M.E. and Fennimore S.A. (1998) Genetic basis for seed dormancy. *Seed Science Research*, 8: 173-182.

Froud-Williams R.J., Drennan D.S.H. and Chancellor R.J. (1984b). The influence of burial and dry-storage upon cyclic changes in dormancy, germination and response to light in seeds of various arable weeds. *New Phytologist* 96, 473-481.

Froud-Williams R.J., Hilton J.R. and Dixon J. (1986) Evidence for endogenous cycle of dormancy in dry stored seeds of *Poa trivialis* L. *New Phytologist* 102:123-131.

Givelberg A., Horowitz M. and Poljakoff-Mayber A. (1984) Germination behavior of *Solanum nigrum* seeds. *Journal of Experimental Botany* 35: 588–598.

Goudey J.S., Saini H.S. and Spencer M.S. (1988) Role of nitrate in regulating germination of *Sinapis arvensis* L. (wild mustard). *Plant, Cell andandandandand Environment* 11:9–12.

Grant D.W., Peters D.P.C., Beck G.K. and Fraleigh H.D. (2003) Influence of an exotic species, *Acroptilon repens* (L.) DC. on seedling emergence and growth of native grasses. *Plant Ecology* 166(2): 157-166.

Grundy A.C. (2003) Predicting weed emergence: a review of approaches and future challenges. *Weed Re**search*** **43:** 1–11.

Grundy A.C., Mead A. and Burston S. (2003) Modelling the emergence response of weed seeds to burial depth: interactions with seed density, weight and shape. *Journal of Applied Ecology* 40: 757–770.

Gu X.Y., Kianian S.F. and Foley M.E. (2004) Multiple loci and epistases control genetic variation for seed dormancy in weedy rice (*Oryza sativa*). *Genetics* 166: 1503–1516.

Gulshan A.B. and Dasti A.A. (2012) Role of soil texture and depths one the emergence of buried weed seeds. *ARPN Journal of Agricultural and Biological Science* 7(4): 223-228.

Guo P. and Al-Khatib K. (2003) Temperature effects on germination and growth of redroot pigweed (*Amaranthus retroflexus*), Palmer amaranth (*A. palmeri*), and common waterhemp (*A. rudis*) *Weed Science* 51(6): 869-875.

Håkansson S. (1983) Seasonal variation in the emergence of annual weeds-an introductory investigation in Sweden. *Weed Research* 23: 313-324.

Handley R.J. and Davy A.J. (2005) Temperature effects on seed maturity and dormancy cycles in an aquatic annual, *Najas marina*, at the edge of its range. *Journal of Ecology* 93: 1185–1193.

Haramoto E. R. and Gallandt E. R. (2005), *Brassica* cover cropping: I. Effects on weed and crop establishment. *Weed Science,* 53(5): 695-701.

Hilhorst H.W.M. (1995) A critical update on seed dormancy. I. Primary dormancy. *Seed Science Research* 5: 61-73.

Hilhorst H.W.M. and Karssen C.M. (1992) Seed dormancy and germination: the role of abscisic acid and gibberellins and the importance of hormone mutants. *Plant Growth Regulation* 11: 225-238.

Hilhorst H.W.M., Smitt A. and Karssen C.M. (1986) Gibberellin-biosynthesis and -sensitivity mediated stimulation of seed germination of *Sisymbrium officinale* by red light and nitrate. *Physiologia Plantarum* 67: 285-90.

Holm L.G., Plucknett D.L., Pancho J.V. and Herberger J.P. (1991) *The World's Worst Weeds. Distribution and Biology*. Honolulu, HI: University Press of Hawaii.

Honek A., Martinkova Z. and Jarosik V. (1999) Annual cycles of germinability and differences between primary and secondary dormancy in buried seeds of *Echinochloa crus-galli*. *Weed Research* 39: 69–79.

Hoyle G.L., Steadman K.J., Daws,M.I. and Adkins S.W. (2008) Pre- and post-harvest influences on seed dormancy status of an Australian Goodeniaceae species, *Goodenia fascicularis*. *Annals of Botany* 102: 93–101.

Huarte R. and Benech-Arnold R.L. (2005) Incubation under fluctuating temperatures reduces mean base water potential for seed germination in several non-cultivated species. *Seed Science Research* 15: 89-97.

Hyakutake H. and Yoshida S. (1993) Latitudinal differentiation in heading photoperiod sensitivity and seed dormancy of *Echinochloa oryzicola*, an obligate weed in flooded rice. Proceeding of the 10th Australian Weeds Conference/14th Asian-Pacific Weed Science Society Conference, 6-10 September 1993, Brisbane, Queensland, Australia, VL (1): 134.

Hyatt L.A., Evans A.S. and Baskin C.C. (1999.)Annual dormancy cycles in *Lesquerella fendleri* (Brassicaceae) seeds stored under both field and laboratory conditions. *Canadian Journal of Botany* 77: 1648-1654.

Jha P., Norsworthy J.K., Riley M.B. and Bridges Jr W. (2010) Annual Changes in Temperature and Light Requirements for Germination of Palmer Amaranth (*Amaranthus palmeri*) Seeds Retrieved from Soil. *Weed Science* 58(4): 426-432.

Juroszek P. and Gerhards R. (2004) Photocontrol of weeds. *Journal of Agronomy and Crop Science* 190: 402–415.

Jursík M., Soukup J., Venclová V. and Holec J. (2003) Seed dormancy and germination of Shaggy soldier (*Galinsoga ciliata* Blake.) and Common lambsquarter (*Chenopodium album* L.). *Plant and Soil Environment* 49(11): 511–518.

Karssen C.M. (1970) Light-promoted germination of the seeds of *Chenopodium album* L. III. Effect of the photoperiod during growth and development of the plants on the dormancy of the produced seeds. *Acta Botanica of Netherland*, 19: 81-94.

Karssen C.M., Brinkhorst-Van Der Swain D.L.C., Breekland A.E. and Koornneef M. (1983) Induction of dormancy during seed development by endogenous abscisic acid: studies on abscisic acid deficient genotypes of *Arabidopsis thaliana* (L.). Heynh. Planta 157: 158- 165.

Kaul M.L.H. and Mangal P.D. (1987) Phenology and germination of crownbeard (*Verbesina encelioides*). *Weed Science* 35(4): 513-518.

Kebreab E. and Murdoch A.J. (1999) A quantitative model for loss of primary dormancy and induction of secondary dormancy in imbibed seeds of *Orobanche* spp. *Journal of Experimental Botany* **50**: 211–219.

Keeley P.E. and Thullen R.J. (1983) Influence of planting date on the growth of black nightshade (*Solanum nigrum*). *Weed Science* 31: 180–184.

Kegode G.O. and Pearce R.B. (1998) Influence of environment during maternal plant growth on dormancy of shattercane (*Sorghum bicolor*) and giant foxtail (*Setaria faberi*) seed. *Weed Science* 46: 322–329.

Kermode A.R. (2005) Role of abscisic acid in seed dormancy. Journal of Plant Growth Regulation 24: 319-344.

Khan A.A. and Karssen C.M. (1980) Induction of secondary dormancy in *Chenopodium bonus-henricus* L. seeds by osmotic and high temperature treatments and its prevention by light and growth regulators. *Plant Physiology* 66(1): 175–181.

Kleemann S.G.L. and Gill G.S. (2012) Seed dormancy and seedling emergence in ripgut brome (*Bromus diandrus* Roth) populations in southern Australia. Weed Science In-Press.

Koger C.H., Reddy K.N. and Poston D.H. (2004) Factors affecting seed germination, seedling emergence, and survival of texasweed (*Caperonia palustris*). *Weed Science* 52:989–995.

Kon K.F., Follas G.B. and James D.E. (2007) Seed dormancy and germination phenology of grass weeds and implications for their control in cereals. New Zealand Plant Protection 60:174-182 (2007).

Koornneef M., Bentsink L. and Hilhorst H. (2002) Seed dormancy and germination. *Current Opinion in Plant Biology* 5: 33–36.

Kremer E. and Lotz L.A.P. (1998) Germination and emergence characteristics of triazine-susceptible and triazine-resistant biotypes of *Solanum nigrum*. *Journal of Applied Ecology* 35: 302–310.

Larina S.Yu. (2008) *Brassica juncea*. In A.N. Afonin, S.L. Greene, N.I. Dzyubenko, A.N. Frolov (eds.). 2008. Interactive Agricultural Ecological Atlas of Russia and Neighboring Countries. Economic Plants and their

Diseases, Pests and Weeds [Online]. Available at: http://www.agroatlas. ru /en/content/weeds/Brassica_juncea/

Larina S.Yu. (2008) *Capsella bursa-pastoris*. In A.N. Afonin, S.L. Greene, N.I. Dzyubenko, A.N. Frolov (eds.). 2008. Interactive Agricultural Ecological Atlas of Russia and Neighboring Countries. Economic Plants and their Diseases, Pests and Weeds [Online]. Available at: http://www.agroatlas.ru/en/content/weeds/Capsella_bursa-pastoris/

Larina S.Yu. (2008) *Cardaria draba*. In A.N. Afonin, S.L. Greene, N.I. Dzyubenko, A.N. Frolov (eds.). 2008. Interactive Agricultural Ecological Atlas of Russia and Neighboring Countries. Economic Plants and their Diseases, Pests and Weeds [Online]. Available at: http://www.agroatlas. ru /en/content/weeds/Cardaria_draba/

Larina S.Yu. (2008) *Chenopodium album*. In A.N. Afonin, S.L. Greene, N.I. Dzyubenko, A.N. Frolov (eds.). 2008. Interactive Agricultural Ecological Atlas of Russia and Neighboring Countries. Economic Plants and their Diseases, Pests and Weeds [Online]. Available at: http://www.agroatlas. ru/ en/content/weeds/Chenopodium_album/

Larina S.Yu. (2008) *Raphanus raphanistrum*. In A.N. Afonin, S.L. Greene, N.I. Dzyubenko, A.N. Frolov (eds.). 2008. Interactive Agricultural Ecological Atlas of Russia and Neighboring Countries. Economic Plants and their Diseases, Pests and Weeds [Online]. Available at: http://www.agroatlas.ru/en/content/weeds/Raphanus_raphanistrum/

Larina S.Yu. (2008) *Sinapis arvensis*. In A.N. Afonin, S.L. Greene, N.I. Dzyubenko, A.N. Frolov (eds.). 2008. Interactive Agricultural Ecological Atlas of Russia and Neighboring Countries. Economic Plants and their Diseases, Pests and Weeds [Online]. Available at: http://www.agroatlas. ru/ en/content/weeds/Sinapis_arvensis/

Larina S.Yu. (2008) *Solanum nigrum*. In A.N. Afonin, S.L. Greene, N.I. Dzyubenko, A.N. Frolov (eds.). 2008. Interactive Agricultural Ecological Atlas of Russia and Neighboring Countries. Economic Plants and their Diseases, Pests and Weeds [Online]. Available at: http://www.agroatlas. ru/ en/content/weeds/Solanum_nigrum/

Larina S.Yu. (2008) *Thlaspi arvense*. In A.N. Afonin, S.L. Greene, N.I. Dzyubenko, A.N. Frolov (eds.). 2008. Interactive Agricultural Ecological Atlas of Russia and Neighboring Countries. Economic Plants and their Diseases, Pests and Weeds [Online]. Available at: http://www.agroatlas. ru/ en/content/weeds/Thlaspi_arvense/

Leblanc M.L., Cloutier D.C., Légère A., Lemieux C., Assémat L., Benoit D.L. and Hamel C. (2002) Effect of the presence or absence of corn on common lambsquarters (*Chenopodium album* L.) and barnyardgrass [*Echinochloa crus-galli* (L.) Beauv.] Emergence. *Weed Technology* 16(3): 638-644.

Leon R.G. and Owen M.D.K. (2003) Regulation of weed seed dormancy through light and temperature interactions. *Weed Science* 51: 752–758.

Leon R.G., Bassham D.C. and Owen M.D.K. (2006) Germination and proteome analyses reveal intraspecific variation in seed dormancy regulation in common waterhemp (*Amaranthus tuberculatus*). Weed Science 54(2): 305-315.

Li Y.P. and Feng Y.L. (2009) Differences in seed morphometric and germination traits of crofton weed (*Eupatorium adenophorum*) from different elevations. Weed Science 57(1) 26-30.

Li X.J., Zhang M., Wei S.H. and Cui H.L. (2011) Influence of environmental factors on seed germination and seedling emergence of yellowtop (*Flaveria bidentis*). 23rd Asian-Pacific Weed Science Society Conference. Volume 1: weed management in a changing world, Cairns, Queensland, Australia, 26-29 September 2011, pp 270-277.

Lu P, Sang W.G. and Ma K.P. (2006) Effects of environmental factors on germination and emergence of Crofton weed (*Eupatorium adenophorum*). Weed Science: May 2006, Vol. 54, No. 3, pp. 452-457.

Luneva N.N. (2008) *Apera spica-venti*. In A.N. Afonin, S.L. Greene, N.I. Dzyubenko, A.N. Frolov (eds.). 2008. Interactive Agricultural Ecological Atlas of Russia and Neighboring Countries. Economic Plants and their Diseases, Pests and Weeds [Online]. Available at: http://www.agroatlas.ru/ en/content/weeds/Apera_spica-venti/

Luneva N.N. (2008) *Avena fatua.* In A.N. Afonin, S.L. Greene, N.I. Dzyubenko, A.N. Frolov (eds.). 2008. Interactive Agricultural Ecological Atlas of Russia and Neighboring Countries. Economic Plants and their Diseases, Pests and Weeds [Online]. Available at: http://www. agroatlas.ru/ en/content/weeds/Avena_fatua/

Luneva N.N. (2008) *Echinochloa crusgalli.* In A.N. Afonin, S.L. Greene, N.I. Dzyubenko, A.N. Frolov (eds.). 2008. Interactive Agricultural Ecological Atlas of Russia and Neighboring Countries. Economic Plants and their Diseases, Pests and Weeds [Online]. Available at: http://www.agroatlas.ru/ en/content/weeds/Echinochloa_crusgalli/

Malik M.S., Norsworthy J.K., Riley M.B. and Bridges Jr. W. (2010) Temperature and light requirements for wild radish (*Raphanus raphanistrum*) germination over a 12-month period following maturation. *Weed Science* 58(2):136-140.

Matilla A., Gallardo M. and Puga-Hermida M.I. (2005) Structural, physiological and molecular aspects of heterogeneity in seeds: a review. *Seed Science Research*, 15, pp 63-76.

Matus-Cadiz M.A. and Hucl P. (2005) Rapid and effective germination methods for overcoming seed dormancy in annual canarygrass. *Crop Science* 45:1696-703.

Mayeux Jr. H.S. (1983) Effects of soil texture and seed placement on emergence of four subshrubs. *Weed Science,* 31(3): 380-384.

Mekenian M. R. and Willemsen R. W. (1975) Germination characteristics of *Raphanus raphanistrum*. I. laboratory studies. Bulletin of the Torrey Botanical Club 102(5): 243-252.

Mennan H. and Ngouajio,M. (2006) Seasonal cycle in germination and seedling emergence of summer and winter populations of catchweed bedstraw (*Galium aparine*) and wild mustard (*Brassica kaber*). *Weed Science* 54: 114-120.

Milberg P. and Andersson L. (1998) Does cold stratification level out differences in seed germinability between populations? *Plant Ecology* 134: 225-234.

Milberg P., Andersson L., Elfverson C. and Regner S. (1996) Germination characteristics of seeds differing in mass. *Seed Science Research* 6: 191–197.

Mumford P.M. (1988) Alleviation and induction of dormancy by temperature in *Impatiens glandulifera* Royle. New Phytologist, 109: 107-110.

Munir J., Dorn L. A., Donohue K. and Schmitt J. (2001) The effect of maternal photoperiod on seasonal dormancy in *Arabidopsis thaliana* (Brassicaceae). *American Journal of Botany* 88: 1240-1249.

Murrumbidgee CMA (Undated) *Best management practices for dryland cropping systems: Wild Radish (Raphanus raphanistrum).*, NSW. URL: http://www.murrumbidgee.cma.nsw.gov.au/downloads/info_sheets/dryland_cropping/Wild_Radish_Revised.pdf - Accessed January 2013.

Mutz J.L. and Scifres C.J. (1975) Soil texture and planting depth influence buffelgrass emergence. *Journal of Range Management* 28(3): 222-224.

Myers M.W., Curran W.S., Vangessel M.J., Majek B.A., Mortensen D.A., Calvin D.D., Karsten H.[D. and Roth G.W. (2005) Effect of soil disturbance on annual weed emergence in the northeastern United States. *Weed Technology* 19:274–282.

Nadtochii I.N. (2008) Ambrosia *Ambrosia artemisiifolia*. In A.N. Afonin, S.L. Greene, N.I. Dzyubenko, A.N. Frolov (eds.). 2008. Interactive Agricultural Ecological Atlas of Russia and Neighboring Countries.

Economic Plants and their Diseases, Pests and Weeds [Online]. Available at: http://www.agroatlas.ru/en/content/weeds/Ambrosia_artemisiifolia/

Nadtochii I.N. (2008) *Anthemis arvensis*. In A.N. Afonin, S.L. Greene, N.I. Dzyubenko, A.N. Frolov (eds.). 2008. Interactive Agricultural Ecological Atlas of Russia and Neighboring Countries. Economic Plants and their Diseases, Pests and Weeds [Online]. Available at: http://www.agroatlas.ru/ en/content/weeds/Anthemis_arvensis/

Nadtochii I.N. (2008) *Artemisia vulgaris*. In A.N. Afonin, S.L. Greene, N.I. Dzyubenko, A.N. Frolov (eds.). 2008. Interactive Agricultural Ecological Atlas of Russia and Neighboring Countries. Economic Plants and their Diseases, Pests and Weeds [Online]. Available at: http://www.agroatlas.ru/ en/content/weeds/Artemisia_vulgaris/

Nadtochii I.N. (2008) *Centaurea cyanus*. In A.N. Afonin, S.L. Greene, N.I. Dzyubenko, A.N. Frolov (eds.). 2008. Interactive Agricultural Ecological Atlas of Russia and Neighboring Countries. Economic Plants and their Diseases, Pests and Weeds [Online]. Available at: http://www.agroatlas.ru/ en/content/weeds/Centaurea_cyanus/

Nadtochii I.N. (2008) *Convolvulus arvensis.* In A.N. Afonin, S.L. Greene, N.I. Dzyubenko, A.N. Frolov (eds.). 2008. Interactive Agricultural Ecological Atlas of Russia and Neighboring Countries. Economic Plants and their Diseases, Pests and Weeds [Online]. Available at: http://www.agroatlas.ru/en/content/weeds/Convolvulus_arvensis/

Nadtochii I.N. (2008) *Helianthus lenticularis.* In A.N. Afonin, S.L. Greene, N.I. Dzyubenko, A.N. Frolov (eds.). 2008. Interactive Agricultural Ecological Atlas of Russia and Neighboring Countries. Economic Plants and their Diseases, Pests and Weeds [Online]. Available at: http://www.agroatlas.ru/en/content/weeds/Helianthus_lenticularis/

Nadtochii I.N. (2008) *Plantago major.* In A.N. Afonin, S.L. Greene, N.I. Dzyubenko, A.N. Frolov (eds.). 2008. Interactive Agricultural Ecological Atlas of Russia and Neighboring Countries. Economic Plants and their Diseases, Pests and Weeds [Online]. Available at: http://www.agroatlas.ru/en/content/weeds/Plantago_major/

Nadtochii I.N. (2008) *Portulaca oleracea.* In A.N. Afonin, S.L. Greene, N.I. Dzyubenko, A.N. Frolov (eds.). 2008. Interactive Agricultural Ecological Atlas of Russia and Neighboring Countries. Economic Plants and their Diseases, Pests and Weeds [Online]. Available at: http://www. agroatlas.ru/ en/content/weeds/Portulaca_oleracea/

Nadtochii I.N. (2008) *Reseda lutea.* In A.N. Afonin, S.L. Greene, N.I. Dzyubenko, A.N. Frolov (eds.). 2008. Interactive Agricultural Ecological

Atlas of Russia and Neighboring Countries. Economic Plants and their Diseases, Pests and Weeds [Online]. Available at: http://www.agroatlas.ru/ en/content/weeds/Reseda_lutea/

Nadtochii I.N. (2008) *Viola tricolor.* In A.N. Afonin, S.L. Greene, N.I. Dzyubenko, A.N. Frolov (eds.). 2008. Interactive Agricultural Ecological Atlas of Russia and Neighboring Countries. Economic Plants and their Diseases, Pests and Weeds [Online]. Available at: http://www.agroatlas.ru/ en/content/weeds/Viola_tricolor/

Nadtochii I.N. (2008) *Xanthium strumarium.* In A.N. Afonin, S.L. Greene, N.I. Dzyubenko, A.N. Frolov (eds.). 2008. Interactive Agricultural Ecological Atlas of Russia and Neighboring Countries. Economic Plants and their Diseases, Pests and Weeds [Online]. Available at: http://www.agroatlas.ru/en/content/weeds/Xanthium_strumarium/

Nikolaeva M.G. (2001) Ecological and physiological aspects of seed dormancy and germination (review of investigations for the last century). *Botanicheskii Zhurnal* 86: 1-14.

Noronha A., Andersson L. and Milberg P. (1997) Rate of change in dormancy level and light requirement in weed seeds during stratification. - *Annals of Botany* 80: 795-801.

Norsworthy J.K. and Oliveira M.J. (2005). Coffee senna (*Cassia occidentalis*) germination and emergence is affected by environmental factors and seeding depth. *Weed Science* 53: 657-662.

Norsworthy J.K. and Oliveira M.J. (2006) Sicklepod (*Senna obtusifolia*) germination and emergence as affected by environmental factors and seeding depth. *Weed Science* 54(5): 903-909.

Nurse R. E. and DiTommaso A. (2005) Corn competition alters the germinability of velvetleaf (*Abutilon theophrasti*) seeds. *Weed Science* 53:479–488.

Ogg A.G. Jr. and Dawson J.H. (1984) Time of emergence of eight weed species. *Weed Science* 32:327–335.

Ogg A.G. Jr. and Rogers B.S. (1989) Taxonomy, distribution, biology, and control of black nightshade (*Solanum nigrum*) and related species in the United States and Canada. *Review of Weed Science* 4:25-58.

Penny G.M. and Neal J.C. (2003) Light, temperature, seed burial, and mulch effects on mulberry weed (*Fatoua villosa*) seed germination. *Weed Technology* 17: 213-218.

Peters N.C.B. (1982) The dormancy of wild oat seed (*Avena fatua* L.) from plants grown under various temperature and soil moisture conditions. *Weed Research* 22: 205–212.

Philippi T. (1993) Bet-hedging germination of desert annuals: variation among populations and maternal effects in *Lepidium lasiocarpum. American Naturalist* 142: 488–507.

Popay A.I., Cox T.I., Ingle A. and Kerr R. (1995) Seasonal emergence of weeds in cultivated soil in New Zealand. *Weed Research* 35: 429-436.

Popay A.I. and Roberts E.H. (1970) Factors involved in the dormancy and germination of *Capsella bursa-pastoris* (L.) Medik. and *Senecio vulgaris* L. *Journal of Ecology* 58(1): 103-22.

Rahman A., James T.K., Mellsop J. and Grbavac N. (2000) Effect of cultivation methods on weed seed distribution and seedling emergence. *New Zealand Plant Protection* 53:28-33.

Ray J., Creamer R., Schroeder J. and Murray L. (2005) Moisture and temperature requirements for london rocket (*Sisymbrium irio*) emergence. *Weed Science*, 53(2): 187-192

Roach D.A. and Wulff R.D. (1987) Maternal effects in plants. *Annual Review of Ecology and Systematics* 18: 209–235.

Roberts H.A. and Boddrell J.E. (1983) Field emergence and temperature requirements for germination in *Solanum sarrachoides* Sendt. *Weed Research* 23: 247–252.

Roberts H.A. and Lockett P.M. (1978) Seed dormancy and field emergence in *Solanum nigrum* L. *Weed Research* 18: 231-241.

Rothlisberger K.L., Hons F.M., Gentry T.J., and Senseman S.A. (2012) Oilseed meal effects on the emergence and survival of crop and weed species. *Applied and Environmental Soil Science* Volume 2012, Article ID 769357, 10 pages.

Sabila M.H., Grey T.L., Webster T.M., Vencill W.K. and Shilling D.G. (2012) Evaluation of factors that influence benghal dayflower (*Commelina benghalensis*) seed germination and emergence. *Weed Science* 60(1): 75-80.

Schonbeck M.W. and Egley G.H. (1980) Redroot pigweed (*Amranthus retroflexus*) seed germination responses to afterripening, temperature, ethylene, and some other environmental factors. Weed Science 28(5): 543-548

Schutte B.J., Regnier E.E. and Harrison S.K. (2012) Seed dormancy and adaptive seedling emergence timing in giant ragweed (*Ambrosia trifida*). Weed Science 60(1): 19-26.

Shem-Tov S. and Fennimore S.A. (2003) Seasonal changes in annual bluegrass (*Poa annua*) germinability and emergence. *Weed Science* 51(5): 690–695.

Silvertown J. and Charlesworth D. (2006) *Introduction to plant population biology*, 4th edn. Blackwell, Oxford.

Singh S. and Singh M. (2009) Effect of temperature, light and pH on germination of twelve weed species. *Indian Journal of Weed Science,* 41:113-126.

Sokolova T.D. (2008) *Amaranthus albus*. In A.N. Afonin, S.L. Greene, N.I. Dzyubenko, A.N. Frolov (eds.). 2008. Interactive Agricultural Ecological Atlas of Russia and Neighboring Countries. Economic Plants and their Diseases, Pests and Weeds [Online]. Available at: http://www.agroatlas.ru/en/content/weeds/Amaranthus_albus/

Sokolova T.D. (2008) *Amaranthus blitoides*. In A.N. Afonin, S.L. Greene, N.I. Dzyubenko, A.N. Frolov (eds.). 2008. Interactive Agricultural Ecological Atlas of Russia and Neighboring Countries. Economic Plants and their Diseases, Pests and Weeds [Online]. Available at: http://www.agroatlas.ru/en/content/weeds/Amaranthus_ blitoides /

Sokolova T.D. , (2008) *Amaranthus retroflexus*. In A.N. Afonin, S.L. Greene, N.I. Dzyubenko, A.N. Frolov (eds.). 2008. Interactive Agricultural Ecological Atlas of Russia and Neighboring Countries. Economic Plants and their Diseases, Pests and Weeds [Online]. Available at: http://www.agroatlas.ru/en/content/weeds/Amaranthus_retroflexus/

Sokolova T.D. (2008) *Sisymbrium altissimum*. In A.N. Afonin, S.L. Greene, N.I. Dzyubenko, A.N. Frolov (eds.). 2008. Interactive Agricultural Ecological Atlas of Russia and Neighboring Countries. Economic Plants and their Diseases, Pests and Weeds [Online]. Available at http://www.agroatlas.ru/en/content/weeds/Sisymbrium_altissimum/

Sokolova T.D. (2008) *Stellaria media*. In A.N. Afonin, S.L. Greene, N.I. Dzyubenko, A.N. Frolov (eds.). 2008. Interactive Agricultural Ecological Atlas of Russia and Neighboring Countries. Economic Plants and their Diseases, Pests and Weeds [Online]. http://www.agroatlas.ru /en/content/weeds/Stellaria_media/

Stanton R., Wu H. and Lemerle D. (2012) Factors affecting silverleaf nightshade (*Solanum elaeagnifolium*) germination. *Weed Science* 60(1): 42-47.

Steckel L.E., Sprague C.L., Stoller E.W. and Wax L.M. (2004) Temperature effects on germination of nine *Amaranthus* species. *Weed Science* 52(2):217-221.

Steadman K.J. and Pritchard H.W. (2003) Germination of *Aesculus hippocastanum* seeds following cold-induced dormancy loss can be described in relation to a temperature-depended reduction in base temperature (T_b) and thermal time,.*New Phytologist* 161: 415–425.

Steadman K.J. (2004) Dormancy release during hydrated storage in *Lolium rigidum* seeds is dependent on temperature, light quality, and hydration status. *Journal of Experimental Botany* 55: 929-937.

Steadman K.J., Ellery A.J., Chapman R., Moore A. and Turner N.C. (2004) Maturation temperature and rainfall influence seed dormancy characteristics of annual ryegrass (*Lolium rigidum*). *Australian Journal of Agricultural Research* 55:1047–1057.

Stoller E.W. and Wax L.M. (1973) Periodicity of germination and emergence of some annual weeds. *Weed Science* 21:574–580.

Strand E. (1989) Studies on seed dormancy in small grain species. II. Wheat. *Norwegian Journal of Agricultural Sciences* 3: 101–115.

Taab A. and Striebig C. J. (2005) A study on wild oat populations in Iran. In: 13th EWRS (European Weed Research Society) Symposium, 20-23 June 2005 Bari/Italy.

Taab A. (2009) Seed dormancy and germination characteristics of *Solanum nigrum* and *S. physalifolium* as affected by temperature, PhD thesis, the Swedish University of Agricultural Sciences, Uppsala, Sweden.

Taab A. and Andersson L. (2009a) Seasonal changes in seed dormancy of *Solanum nigrum* and *Solanum physalifolium*. *Weed Research* 49:90–97.

Taab A. and Andersson L. (2009b) Seed dormancy dynamics and germination characteristics of *Solanum nigrum*. *Weed Research* 49 (5): 490-498.

Taab A. and Andersson L. (2009c) Primary dormancy and seedling emergence of black nightshade (*Solanum nigrum*) and hairy nightshade (*S. physalifolium*). *Weed Science* 57:526–532.

Taab A., Akbari M. and Anbari S. (2011a) Seedling emergence of *Amaranthus blitoides* affected by soil burial depth. European Weed Research Society Joint Workshop, 4-8 September 2011, Huesca, Spain.

Taab A., Mottaghi Shahpar M. and Anbari S. (2011b) Effect of soil burial depth on seedling emergence of *Amaranthus retroflexus*. European Weed Research Society Joint Workshop, 4-8 September 2011, Huesca, Spain.

Thompson K. and Grime J.P. (1983) A comparative study of germination responses to diurnally-fluctuating temperatures. *Journal of Applied Ecology* 20: 141–56.

Toorop P.E., Cuerva R.C., Begg., Locardi B., Squire G.R. and Iannetta P.P. (2012) Co-adaptation of seed dormancy and flowering time in the arable weed *Capsella bursa-pastoris* (shepherd's purse). *Annals of Botany* 109(2): 481-489.

Totterdell S. and Roberts E.H. (1979) Effects of low temperatures on the loss of innate dormancy and the development of induced dormancy in seeds of *Rumex obtusifolius* L. and *Rumex crispus* L. *Plant, Cell and Environment* 2: 131–137.

Vleeshouwers L.M. (1997a) Modeling the effect of temperature, soil penetration resistance, burial depth and seed weight on pre-emergence growth of weeds. *Annals of Botany* 79: 553-563.

Vleeshouwers L.M. (1997b) *Modelling weed emergence pattern.* Wageningen Agricultural University, The Netherlands.

Vleeshouwers L.M. and Bouwmeester H.J. (2001) A simulation model for seasonal changes in dormancy and germination of seeds. *Seed Science Research* 11: 77–92.

Vleeshouwers L.M. and Kropff M.J. (2000) Modelling field emergence patterns in arable weeds. *New Phytologist* 148: 445–457.

Watson C.E. Jr. and Watson V.H. (1982) Nitrogen and date of defoliation effects on seed yield and seed quality of tall fescue. *Agronomy Journal* 74, 891–893.

Wei S.H., Zhang C.X., Li X.J., Cui H.L., Huang H.J., Sui B.F., Meng Q.H. and Zhang H.J. (2009) Factors affecting buffalobur (*Solanum rostratum*) seed germination and seedling emergence. *Weed Science* 57: 521–525.

Widderick M.J., Walker S.R. and McLean A.R. (2010) Manipulating the seed bank to manage herbicide-resistant weeds. *Seventeenth Australasian Weeds Conference,* 26-30 September 2010, Christchurch, New Zealand, 262-265.

Willemsen R.W. (1975) Effect of stratification temperature and germination temperature on germination and the induction of secondary dormancy in common ragweed seeds. *American Journal of Botany* 62: 1–5

Williams II M. M., Schutte B. J. and So Y. F. (*2012*) Maternal Corn Environment Influences Wild-Proso Millet (*Panicum miliaceum*) Seed Characteristics. *Weed Science* 60(1): 69-74.

Yakutkin V.I. (2008) *Orobanche cumana*. In A.N. Afonin, S.L. Greene, N.I. Dzyubenko, A.N. Frolov (eds.). 2008. Interactive Agricultural Ecological Atlas of Russia and Neighboring Countries. Economic Plants and their Diseases, Pests and Weeds [Online]. Available at http://www.agroatlas.ru/en/content/weeds/Orobanche_cumana/

Yamasue V., Nakatini K. and Kusanagi T. (1993) Latitudinal differentiation in heading photoperiod sensitivity and seed dormancy of *Echinochloa oryzicola*, an obligate weed in flooded rice. Proceeding of the 10th Australian Weeds Conference/14th Asian-Pacific Weed Science Society

Conference, 6-10 September 1993, Brisbane, Queensland, Australia, VL (1) pp 395-399.

Young K. and Cousens R. (1999) Factors affecting the germination and emergence of wild radish (*Raphanus raphanistrum*) and their effect on management options. Proceeding of the 12th Australian Weeds Conference, Hobart, Tasmania, Australia, 12-16 September 1999, 179 - 182

Zhou J., Deckard E.L. and Ahrens W.H. (2005a) Factors affecting germination of hairy nightshade (*Solanum sarrachoides*) seeds. *Weed Science* 53(1) 41-45.

Zhou J., Deckard E.L. and Messersmith C.G. (2005b) Factors affecting eastern black nightshade (*Solanum ptycanthum*) seed germination. *Weed Science* 53(5): 651-656.

Zhou J., Tao B., Deckard E.L., and Messersmith C.G. (2006) Garden huckleberry (*Solanum melanocerasium*) germination, seed survival, and response to herbicides. *Weed Science*: 54(3): 478-483.

Zimmerman C. A. (1977) A comparison of breeding systems and seed physiologies in three species of *Portulaca* L. *Ecology* 58:860–868.

In: Weeds and Their Ecological Functions
Editor: Alireza Taab
ISBN: 978-1-62808-332-3

Chapter 2

SEEDLING EMERGENCE CHARACTERISTICS OF *CARTHAMUS OXYACANTHA* AND *VACCARIA PYRAMIDATA* IN THE SOIL PROFILE: WINTER ANNUAL SPECIES

Alireza Taab[1*], Saghi Anbari[2], Mehrangiz Akbari[1] and Maryam Mottaghi Shahpar[1]

[1]Department of Agronomy and Plant Breeding,
Faculty of Agriculture, Ilam University, Ilam, Iran
[2]Department of Crop Production Ecology,
Swedish University of Agricultural Sciences, Uppsala, Sweden

ABSTRACT

Carthamus oxyacantha and *Vaccaria pyramidata* are winter annual weed species frequently found in rainfed crops e.g. wheat and barley. Seedling emergence patterns of *C. oxyacantha* and *V. pyramidata* were studied with seeds buried at soil depths of 0, 2, 4, 6, 8, and 10 cm in outdoor pots during two years. The seedling emergence of *C. oxyacantha* mainly took place during February to March after which it was negligible and sporadic. The seedling emergence slightly delayed with increasing of

* Correspondence: Alireza Taab, Ph.D., Department of Agronomy and Plant Breeding, Faculty of Agriculture, Ilam University, 69315-516 Ilam, Iran. Tel: (+98) 918 6149226; Fax: (+98) 841 2227015; E-mail: a.taab@mail.ilam.ac.ir.

soil depth. Start of seedling emergence was coincided with a daily mean soil temperature of 5°C at all soil depths. The main flush of seedling emerged when the minimum and maximum soil temperatures were between 4°C to 20°C, respectively. In the first year of the experiment, total seedling emergence was 45.5 % for seeds placed at soil surface, it increased in seeds buried at 2 cm depth (88 %) followed by a decrease at the rest of the depths. However, total seedling emergence did not show a consistent pattern in the second year. The seedling emergence of *V. pyramidata* was also took place from February to March with fewer and sporadic emergences after this time. A minimum daily soil temperature of ca 4°C was found necessary for start of seedling emergence. Most of the seedlings emerged when the minimum and maximum soil temperature were between ca 3 and 14°C, respectively. Total seedling emergence was low for seeds placed at the soil surface in both years (22% in 2010 and 18.5% in 2011); and it was increased at the depth of 2 cm (% 59.5 in 2010 and 51.3% in 2011) followed by a decrease at rest of the depths in 2010. Although large numbers of seedlings of both species may emerge over a short period of time as a main seedling flushes, the sporadic seedling emergences over a relatively extended period enable them to escape weed control operation and maintain problematic in the cultivation.

1. INTRODUCTION

Carthamus oxyacantha Bieb. (wild safflower) and *Vaccaria pyramidata* Medik. (cowcockle) are winter annual troublesome weed species found in many countries and in many crops e.g. wheat, barley, chickpea and lentil. *Carthamus oxyacantha* is found in plains and mountain ranges of Iran in both disturbed and undisturbed areas and due to its spiny-leaved (Bassiri and Rouhani, 1976), it is not readily grazed by livestock. It is, therefore, a prevailing species and growing problem especially in low input agriculture e.g. rainfed crops, where weed control measures are limited. *Vaccaria pyramidata* is also frequently found as a weed in rainfed grains. It is mainly a constant problem in the cultivations due to its large number of seed production. Therefore, the control of these species is of high importance.

Information on seedling emergence characteristics of a weed species helps to improve prediction of emergence timing of the species in a field condition (Taab and Andersson, 2009a). Data on emergence ability in response to burial depths is crucial to develop predictive tools (Forcella *et al.*, 2000) to optimize the timing of weed control measures (Grundy, 2003). Soil disturbances can be adapted to manage weed seed banks. For instance, deeply buried seeds may not be able to emerge or zero tillage may promote greater emergence of species that require light for germination (Chauhan and Johnson, 2009).

Emergence timing of a weed species may affected by several factors. For example, seed dormancy and germination characteristics, soil temperature, moisture and depths of burial in the soil. Emergence studies are usually conducted by recording seedlings of weed species emerging from natural weed infestation (Leblanc *et al.*, 2004) or by sowing seeds at specific soil depths (Weaver *et al.*, 1988). The number of seedlings and rate of seedling emergence was shown to decrease with increasing depth of burial in many species (Benvenuti *et al.*, 2001). Grundy *et al.*, (1996 and 2003a) found that total seedling emergence of *Chenopodium album* substantially declined with increasing depth of burial. Also emergence inhibited at the soil surface.

Environmental factor required for seed germination (i.e. temperature, moisture and light) may vary within the soil profile (Harper, 1977; Stoller and Wax, 1973). The variation in these environmental factors found to affect fate of seeds in the soil (Taylorson, 1970; Stoller and Wax, 1974). Less than 1% of light penetrates only few millimeters into soil (Woolley and Stoller, 1978). Temperature was found to be the major factor affecting weed seedling emergence pattern (Henson, 1970; Blackshaw, 1992; Leblanc *et al.*, 2002).

Pre-emergence treatments such as stratification and scarification were shown to improve the germination of *C. oxyacantha* seeds. However, stratification was more efficient to increase the germination percentages. Thus, an after-ripening period was found to be required to reduce seed dormancy in this species (Bassiri and Rouhani, 1976). Tanveer *et al.*, (2012) reported that seedling emergence in *C. oxyacantha* was decreased with increasing sowing depth and maximum seedlings emerged at soil surface under laboratory condition. Seeds of *V. pyramidata* are capable of survival for a long time (Steiner and Ruckenbauer, 1995). Freshly harvested seeds of *V. pyramidata* showed dormancy that can be reduced under after-ripening condition e.g. at 20°C (Hecker and Köhler, 1979; Hecker *et al.*, 1982). However, at higher temperatures e.g. 30°C, the dormancy could be induced (Hecker and Köhler, 1979). Therefore, both species are likely to present some kind of physiological dormancy (Baskin and Baskin, 2004) that may have implication for regulation of germination and seedling emergence timing of the species (Taab and Andersson, 2009a). However, to our knowledge information on seedling emergence patterns of both species is lacking among the literatures.

There is a progressive demand to reduce weed control inputs e.g. herbicides, due to environmental hazards and cost of crop production. To do so, optimizing the timing of weed control measures is of important value. Therefore, study is needed to understand factors controlling the time and extent of seedling emergence in the field. The objective of this study was to determine the effect of seed burial depths on seedling emergence pattern of *C. oxyacantha* and *V. pyramidata* under field condition.

2. MATERIAL AND METHODS

2.1. Seed Collection

Seeds were collected in early January 2010 for the experiments conducted in 2010 and 2011 from plants growing in a research station (33° 38' 17" North, 46° 25' 50" East) in Ilam, Iran. The seeds were cleaned and kept at ca 5°C until use.

2.2. Sowing Seeds in Pots Outdoors

In mid January 2010 numbers of pots (18-cm diameter, 18-cm height) were prepared for sowing the seeds. The bottom of the pots was open with several holes to ensure similar moisture conditions to the outside pots. The pot were filled and packed with clay soil (clay: 42%; silt: 24% and sand 34%, PH: 7.85 and organic carbon: 1.16%) to nominated depths in each pots so that the surface was measured from 5 cm below the rim. The soil used was extracted from the deep layer of soil (>30 cm; to ensure lack of weed seed infestation) from an experimental field and it was sieved and bulked under plastic film cover for 4 months. The pots were then arranged in a row; each row represented one block, and sunk into the ground with their rims protruding 5 cm above the ground level.

100 seeds were evenly distributed in each pot at 6 soil depths of 0, 2, 4, 6, 8, and 10 cm. The pots were then filled with the soil to 5 cm below the rim. The pots with seeds placed at soil surface were covered with a fine mesh to prevent any possible seed predation. The experimental design was *split-block* (*strip-plot*) consisting of six treatments (seed burial depths) with four replicates. The experiment was repeated in mid January (12/13th) 2011 using a clay soil (clay: 40%; silt: 26% and sand 34%, PH: 7.60 and organic carbon: 3.6%) as mentioned above.

2.3. Soil Temperature

Soil temperature was recorded at all depths during the experiment using temperature loggers (iButton 1-Wire Thermochron DS1921G-F5, Dallas Semiconductor, Inc. USA).

2.4. Seedling Emergence

The pots were surveyed for emergence regularly. The emerged seedlings were counted and removed until the end of the season when no more seedlings were observed. The pots were watered thoroughly when needed using a hand sprinkler during the experiment. The care was taken to avoid any soil disturbances.

2.5. Statistical Analysis

The analysis was conducted to test the differences between years, species, burial depths, and interaction between these factors using the GENMOD procedure (SAS Institute Inc. 2001). The proportion of emerged seedlings was tested as a categorical variable, assuming binomial distribution and logit link function (Table 1). We used a robust method (GEE) to compensate for overdispersion (Diggle *et al.* 2002). Standard error of mean (SEM) was calculated using the LSmeans statement with pdiff option in the GENMOD procedure with account for the assumed binomial distribution. SEMs were then recalculated using back transformation of estimates from the logit scale to the original scale by applying the inverse link function.

The lag time from sowing seeds to start seedling emergence, for all depths in 2010 and 2011, was calculated using the following equation (duration of seedling emergence as independent variable and cumulative emergence as dependent variable):

$$y=(1-\exp(-(a*(x-b)))^c)*d \quad (1)$$

where a is scaling parameter, x is time (day), b is lag time of emergence, c is shape parameter, and d is the final emergence percentage. The calculated lag time of emergence for each depth averaged over both years. To quantify the effect of soil depth on emergence timing, a linear regression equation was fitted to the calculated average lag time of emergence; soil sowing depth as independent variable and average lag time of emergence as dependent variable.

3. Results

3.1. *Carthamus Oxyacantha*

In 2010, seeds of *C. oxyacantha* placed at soil surfaces started to emerge in mid February and continued for a week after which the emergence was declined. This was followed by few sporadic seedling emergences until rest of the experiment. Seeds buried at the depths of 2 to 8 cm emerged after mid February to early March after which the emergence was negligible. The seedling emergence was delayed with depth (Figure 1). For seeds placed at the depth of 10 cm, the emergence was low and sporadic.

In 2011, seeds placed at soil surface started to emerge after mid February with a peak in early April. Seeds buried at the depth of 2 cm started to emerge after mid February with a peak in late February, continued sporadically and declined in the end of March. Seeds at the depth of 4 cm had emergence and a peak in late February and emergence stopped in late April. Seeds at the depth of 6 cm had main flushes of emergence in late February to end of March. Seeds at the depth of 8 cm had main flush of mergence from late February to end of March with a peak in late March. Seeds at the depth of 10 cm had a main flush of emergence from mid to end of March (Figure 2).

Seedling started to emerge when the daily mean soil temperature reached ca 5°C at all soil depths. Most of the seedling emergences took place when minimum and maximum soil temperatures were between 4°C to 20°C, at the depths of 2 to 10 cm. Seedling emergence was declined in end of March when the daily mean soil temperature was between 10°C to 14°C in the preceding weeks, indicating induction of secondary dormancy. The minimum soil temperature was lower at soil surface in compare with depths of 2 to 10 cm. This might be the reason for continued seedling emergence until early April in seeds placed at soil surface (Figure 2).

Record of temperature showed that the daily mean temperature is almost the same at all soil depths. However, fluctuation in temperature decreased with depth (Figure 2).

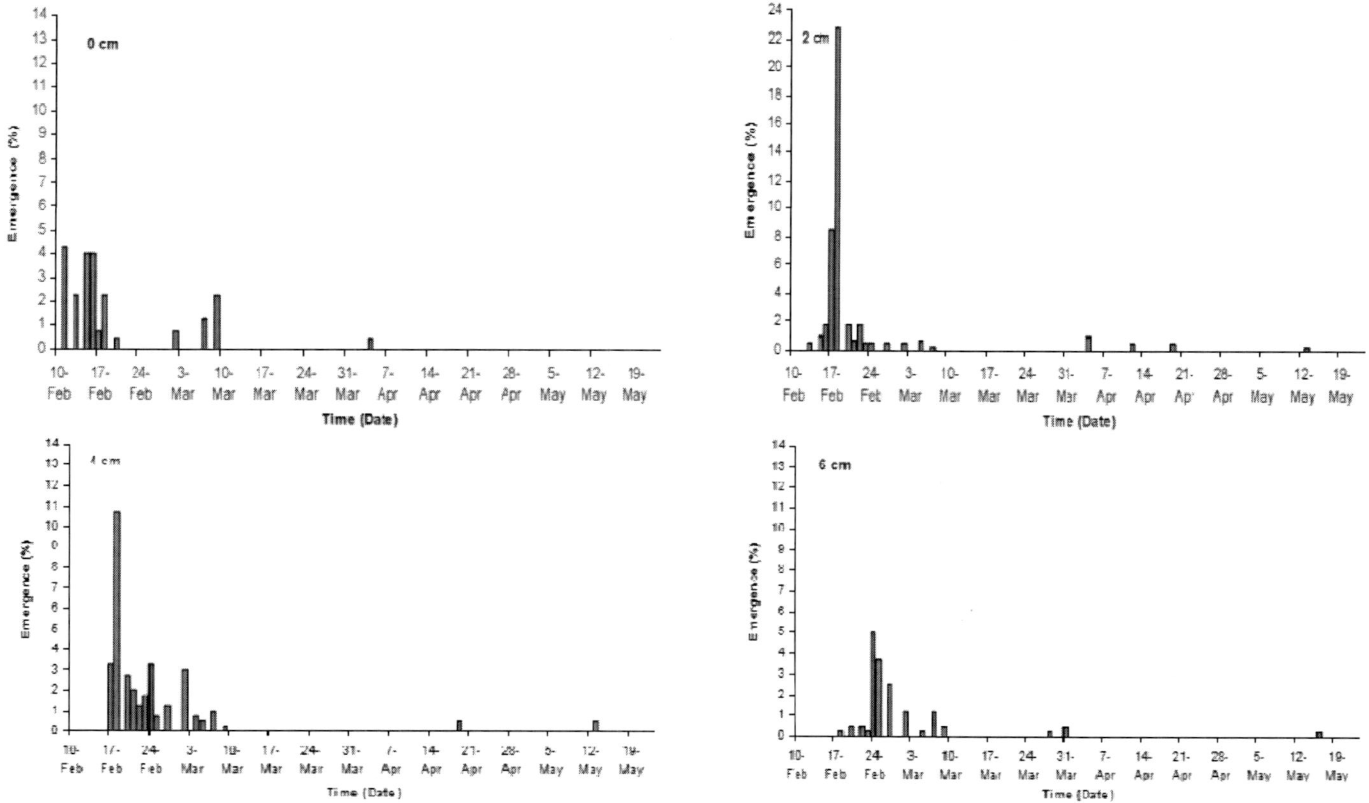

Figure 1. (Continued)

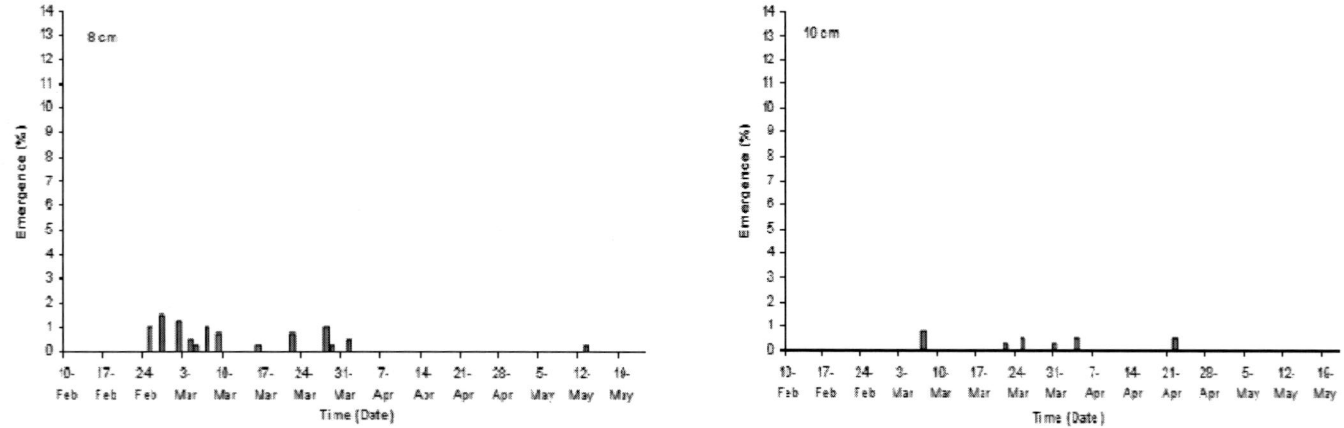

Figure 1. Seedling emergence pattern of *Carthamus oxyacantha* from burial depths of 0 (seeds sown on the soil surface), 2, 4, 6, 8, and 10 cm in 2010.

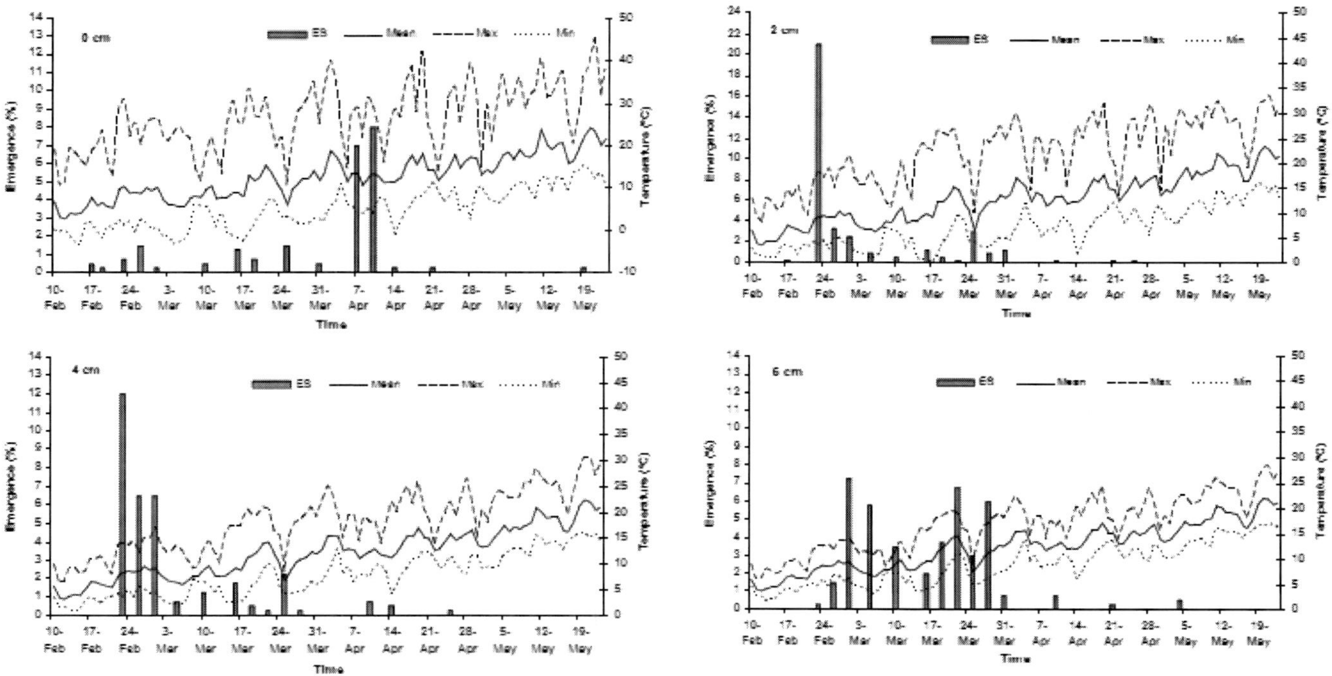

Figure 2. (Continued)

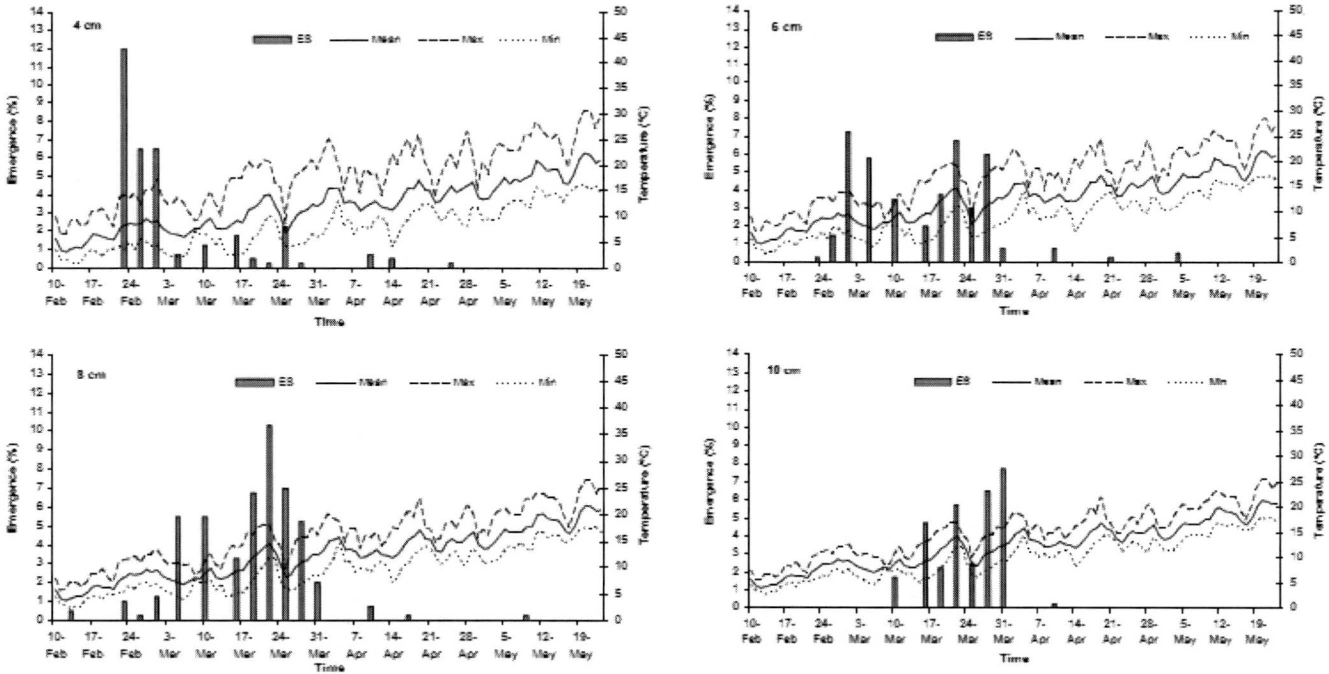

Figure 2. Seedling emergence pattern of *Carthamus oxyacantha* from burial depths of 0 (seeds sown on the soil surface), 2, 4, 6, 8, and 10 cm. ES: emerged seedlings; Mean, Max, and Min: mean, maximum, and minimum soil temperatures, respectively, measured at the burial depths in the pots outdoors in 2011.

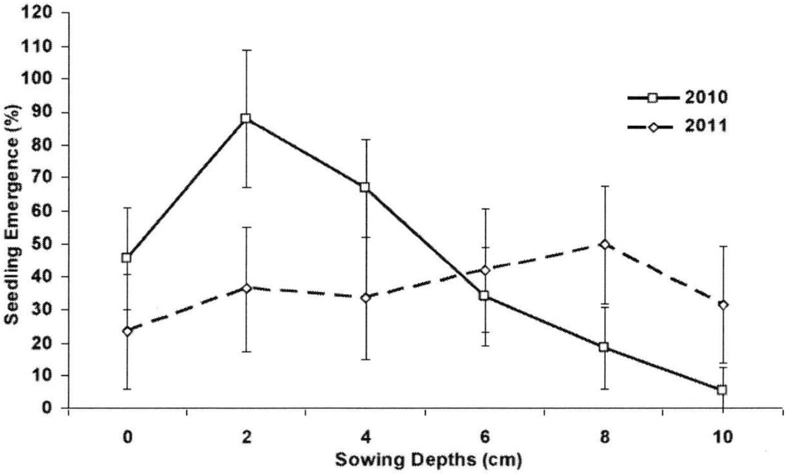

Figure 3. Total number of seedling of *Carthamus oxyacantha* emerging from different soil depths. Vertical bars represent SEM.

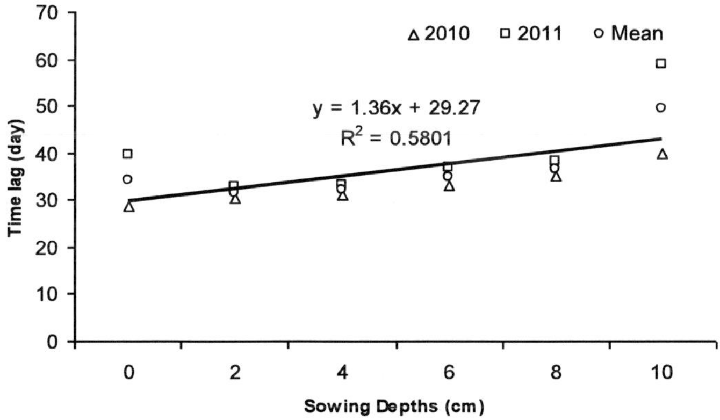

Figure 4. Time lag between sowing date and seedling emergence time of *Carthamus oxyacantha* from different soil depths. Points: calculated and line: predicted time lags.

Total seedling emergence in 2010 was low for seeds at soil surface (%45.5), increased at the depth of 2 cm (%88) followed by decrease at rest of the depths and reached its minimum at the depth of 10 cm (5.5%). In 2011, the total seedling emergence was 23.5% for seeds at the soil surface and it was between 31.5% and 49.8% at the rest of the depths (Figure 3).

According to the regression model the time from sowing seeds to start of seedling emergence is delayed linearly by increasing of soil depth (Figure 4). For instance, seeds at 0 (soil surface), 2, 4, 6, 8, and 10 cm depth may emerge 29.3, 32, 34.7, 37.4, 40.2, and 42.9 days after sowing, respectively.

3.2. *Vaccaria Pyramidata*

In 2010, seeds of *V. pyramidata* had a main flush of seedling emergence from mid to end of February at all depths with few days delay with depth. After this period, seedling emergence was low and sporadic (Figure 5).

In 2011, the seedling emergence was low in seeds placed at the soil surface and started sporadically from mid February to early May. Seeds at the depths of 2 to 10 cm had major flushes of seedling emergence from mid February to early March with fewer number of seedlings emerged after this period (Figure 6).

At all depths, seedlings started to emerge when the daily mean soil temperature reached above ca 4°C. Most of the seedlings emerged when minimum and maximum soil temperatures were between ca 3 and 14°C, respectively, at the depths of 2 to 10 cm. The seedling emergence was tailed off in end of March when the daily mean soil temperature was above 10°C in the preceding week, indicating induction of secondary dormancy. However, the minimum temperature at the soil surface was lower during the same period (Figure 6).

Total seedling emergence was low for seeds placed at the soil surface (22% in 2010 and 18.5% in 2011) and it was increased at the depth of 2 cm (% 59.5 in 2010 and 51.3% in 2011) in both years. This was followed by a decrease at rest of the depths and reached its minimum at the depth of 10 cm (30%) in 2010. However, there was no significant decrease in total emergence in rest of the depths in 2011 (Figure 7).

According to the regression model the start of seedling emergence after sowing is delayed by soil depth linearly (Figure 8). For example, seeds at 0 (soil surface), 2, 4, 6, 8, and 10 cm depth may emerge 30.5, 32, 33.4, 34.9, 36.4, and 37.8 days after sowing, respectively.

Table 1. Analysis of deviance for seedling emergence of *Carthamus oxyacantha* and *Vaccaria pyramidata* from soil depths in 2010 and 2011

Source	DF	Chi-Square	P-value*
Year	1	0.25	0.6196
Species	1	2.59	0.1072
Year*Species	1	0.05	0.8279
Depth	5	23.29	0.0003
Year* Depth	5	21.31	0.0007
Species*Depth	5	13.14	0.0221
Year*Species*Depth	5	14.43	0.0131

* $P< 0.05$ is used as criterion for significance.

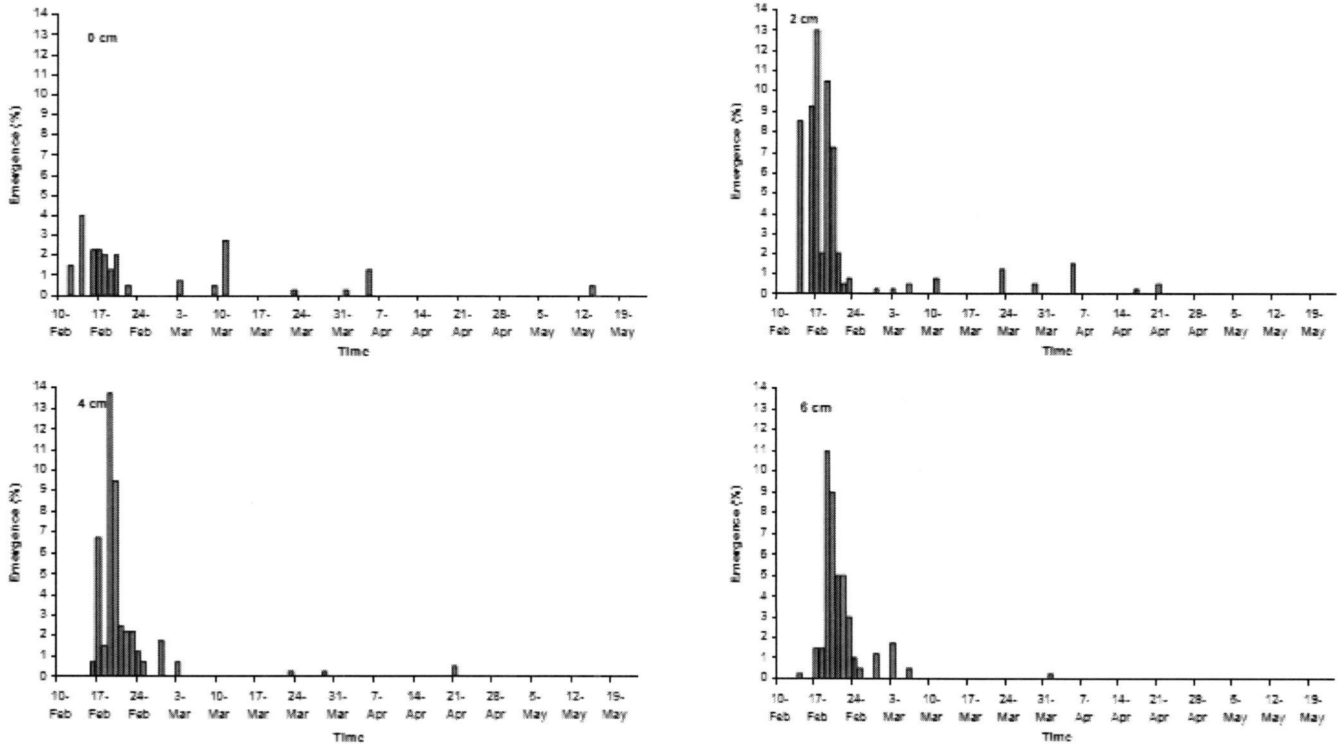

Figure 5. (Continued)

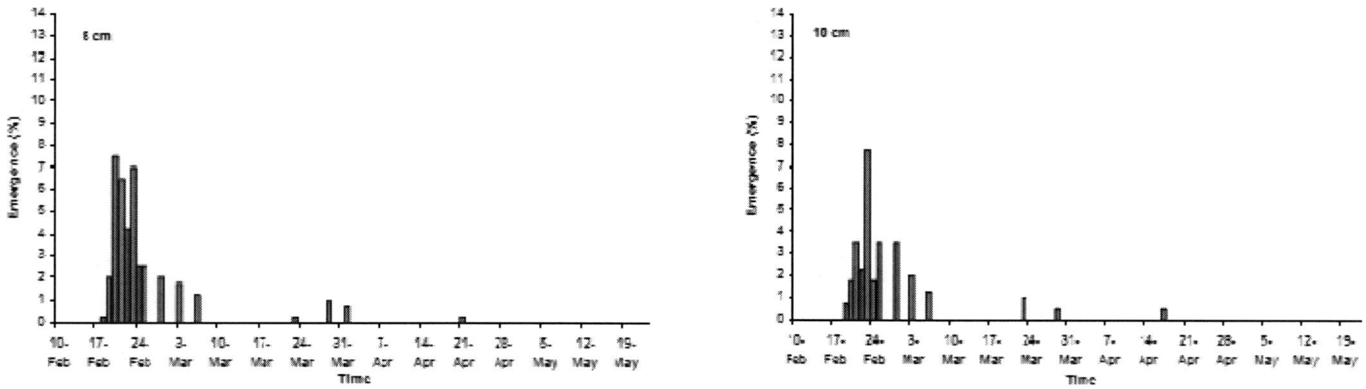

Figure 5. Seedling emergence pattern of *Vaccaria pyramidata* from burial depths of 0 (seeds sown on the soil surface), 2, 4, 6, 8, and 10 cm in 2010.

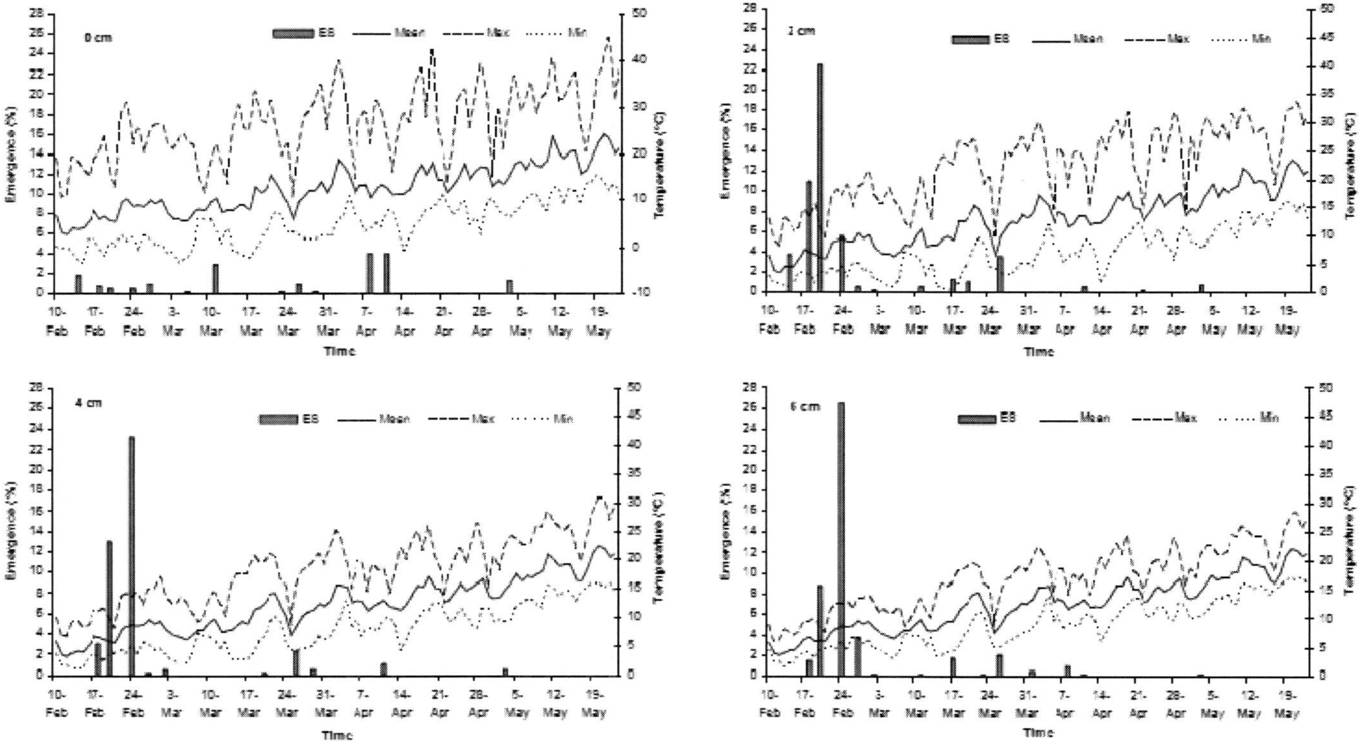

Figure 6. (Continued)

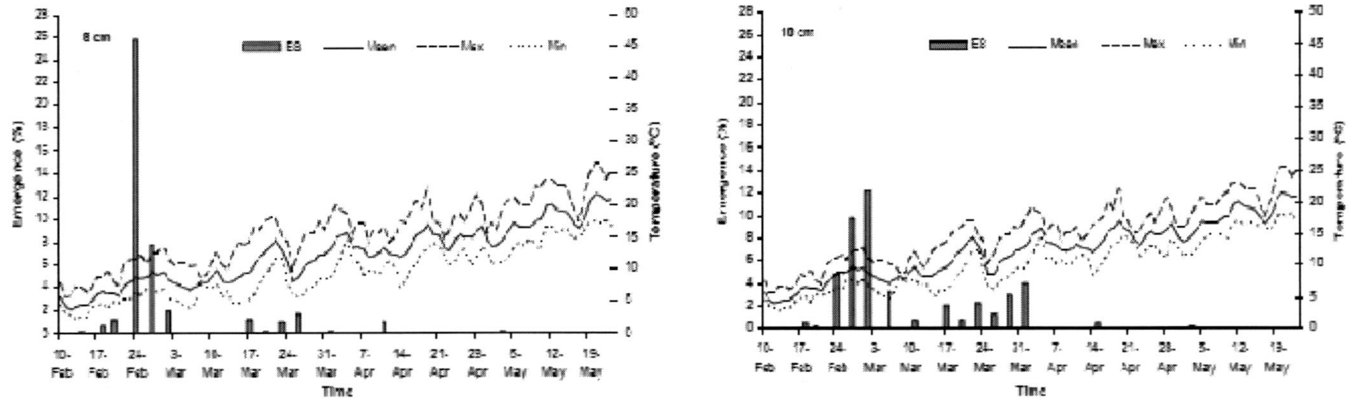

Figure 6. Seedling emergence pattern of *Vaccaria pyramidata* from burial depths of 0 (seeds sown on the soil surface), 2, 4, 6, 8, and 10 cm. ES: emerged seedlings; Mean, Max, and Min: mean, maximum, and minimum soil temperatures, respectively, measured at the burial depths in the pots outdoors in 2011.

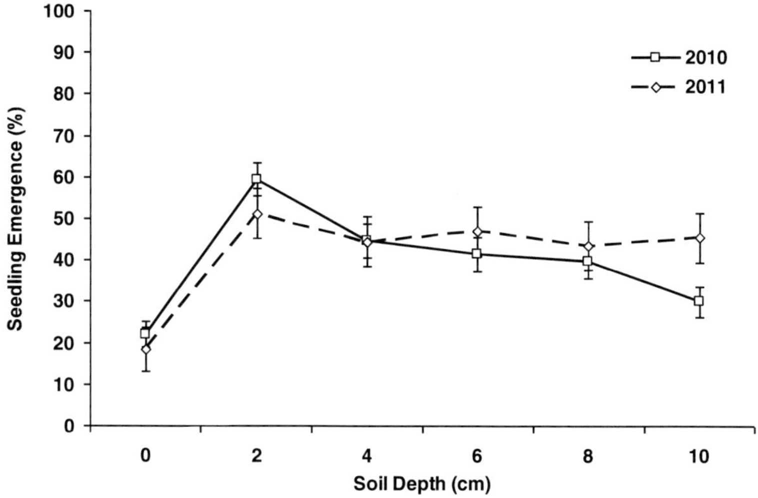

Figure 7. Total number of seedling of *Vaccaria pyramidata* emerging from different soil depths. Vertical bars represent SEM.

4. Discussion

Seedling emergence patterns of *C. oxyacantha* and *V. pyramidata* were affected by soil sowing depth. The emergence of both species started in mid February in seeds placed at soil surface in 2010 and 2011. It was gradually delayed at the depths of 2 to 10 cm by increasing sowing depth (Figure 1, 2, 5 and 6). Total seedling emergence of both species was significantly affected by soil depth (Table 1).

Record of soil temperature showed that a minimum temperature of ca 4°C to 5°C is needed for seedling emergence to start in both species. The daily mean soil temperature was almost the same at all depths. It might be that seedlings emerged at the deeper soil depths need longer time to reach the soil surface in compare to that of seedlings at shallower depths or at soil surface (Figure 4 and 8). For example seedlings of *C. oxyacantha* at 0 (soil surface), may emerge 29.3 after sowing while seedlings at the depth of 10 cm may emerge 42.9 days after sowing. For seedlings of *V. pyramidata*, at 0 (soil surface) may emerge 30.5 days after sowing and seedlings at 10 cm depth may emerge 37.8 days after sowing.

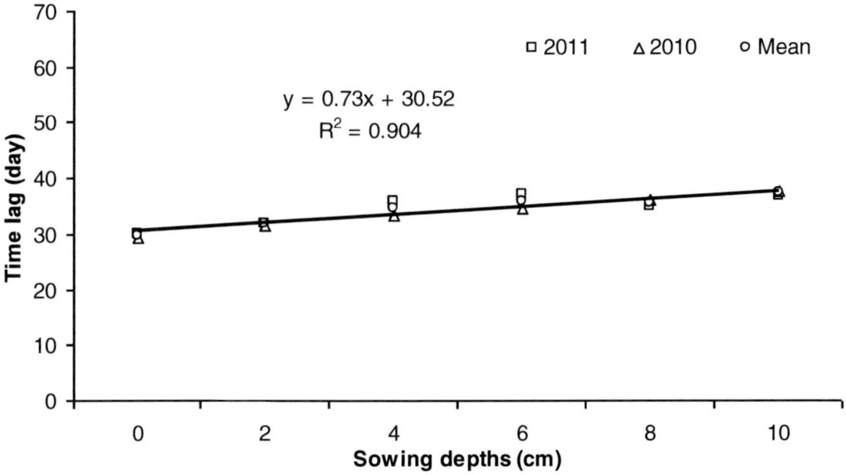

Figure 8. Time lag between sowing date and seedling emergence time of *Vaccaria pyramidata* from different soil depths. Points: calculated and line: predicted time lags.

Comparatively, seedling emergence of *C. oxyacantha* is more delayed by soil depth than that of *V. pyramidata*. Seeds of *C. oxyacantha* are larger than seeds of *V. pyramidata*. As a general rule, larger seeds are expected to have more food reserve to support seed germination and seedling emergence. However, it may take a longer time to imbibe enough water required for bioilogical process before germination to take place in larger-seeded species (Forcella et al., 2000). This could be a reason for slightly more delayed seedling emergence in *C. oxyacantha* than *V. pyramidata*.

In *C. oxyacantha*, the emergence was continued from mid February and tailed off in early March in 2010 at the depths of 0 to 8 cm, after which the seedling emergence was sporadic and negligible. The emergence at the depth of 10 cm was also sporadic and the lowest of all. In 2011, the emergence was also continued from mid February at the depths of 2 to 10 and declined sharply in end of March except for seeds at soil surface that continued to emerge until early April (Figure 1 and 2). In *V. pyramidata*, the seedling emergence was continued during February and tailed off in end of February to beginning of March in 2010, after which emergence was low and sporadic. The same pattern of emergence observed for seeds at the depths of 2 to 10 cm in 2011. However, for seeds at soil surface the emergence was low and extended from February until March (Figure 5 and 6). The results suggest that although large number of seedlings in both species may emerge over a short period as the main flushes, smaller numbers of seedling can emerge over an extended period e.g. from February to May. Seedling emergence over an extended period is known to contribute in the seriousness of a weed species in agricultural conditions. This characteristic at least enables portions of the emerged seedlings to escape weed control operations or climatic extremes and reproduce to guarantee the future weed infestation (Taab and Andersson, 2009a). Therefore, this characteristic could be of important adaptation values for a weed species to survive in the cultivation.

Delayed peak of emergence in seeds of both species at the soil surface, in 2011, could be due to lower minimum soil temperature at soil surface (Figure 2 and 6). According to the recorded temperatures, there were greater temperature fluctuations at soil surface and the fluctuation decreased with soil depth. Saatkamp *et al.*, (2011) reported that seed germination of *V. hispanica* in light and at fluctuating temperature was lower than that of in darkness and at constant temperature. Bassiri *et al.*, (1975) reported that chilling of the seeds of *C. oxyacantha* for a month at 0 °C reduces its seedling emergence. During February and early March 2011, minimum soil temperature was close or below zero at soil surface for most of the time.

In both species, seedling emergence was started in February (at ca 4-5 ºC), increased by increase of soil temperature e.g. ca 10ºC, followed by declined emergence by further increase of soil temperature e.g. daily mean ca 15ºC and maximum temperature above 25 ºC (Figure 2 and 6). Bassiri et al., (1975) studying the germination of *C. oxyacantha* seeds at the temperature ranging form 5 to 30 °C, reported that germination was greatly reduced at the two extreme temperatures of 5 and 30 °C. The optimum temperatures for germination were between 15 and 20 °C. Tanveer et al., (2012) reported that seed germination in *C. oxyacantha* increased at 15ºC, reached its maximum at 20 and 25ºC and decreased at 30ºC. Induction of secondary dormancy at higher temperatures e.g. 30ºC, has been reported in after ripened seeds of *V. pyramidata* (Hecker and Köhler, 1979). Taab and Anderson (2009) showed reduction of primary dormancy followed by induction of secondary dormancy by gradual increasing of temperature from ca 5 to 40°C and at constant temperatures of 15 and 20 °C in stratified seeds of *Solanum nigrum*. In the present study, we concluded that seeds with reduced dormancy have been able to germinate when the soil temperature increased in the favorable range for their germination e.g. in February. However, further increase of temperature e.g. in March and afterward, might cause induction of secondary dormancy.

There is an ecological importance for subsequent survival of seedlings and reproduction of a species with control of germination and emergence timing by dormancy induction (Bewley, 1997). Seedlings that emerge too late in the season may face to sever competition with crop stands. Moreover, in case of winter annual species, they may not be able to complete their life cycle before crop harvest i.e. in wheat. They also may face to sever drought in late spring or early summer during the vegetative growth. Both species showed a relatively similar pattern in total seedling emergence in terms of response to burial depths (Figure 3 and 7). Total seedling emergence was low in seeds of both species placed at the soil surface in both years. It was increased at the depth of 2 cm in *C. oxyacantha* in 2010 and in *V. pyramidata* in 2010 and 2011. Seedling emergence was declined with further increasing of depth in freshly harvested seeds at rest of the depths in 2010. However, in seeds stored at ca 5°C for a year and used in the experiment in 2011, emergence was not declined with depth. The difference in emergence patterns between fresh seeds and stored seeds is likely to be associated with dormancy status of the seeds. Environmental factor required for seed germination (i.e. temperature, moisture and light) may vary within the soil profile (Stoller and Wax, 1973; Harper, 1977). For example, temperature fluctuation is decreased with soil depth (Figure 2 and 6). Seed response to fluctuations in temperature for germination is known as a depth-sensing mechanism (Koller, 1972; Thompson and Grime, 1983).

Dormant seeds may need a period of time conditioning at proper temperature regimes to come out of dormancy and subsequently germinate (Bouwmeester and Karssen 1993). A period of after ripening during summer is required for seeds of winter annual species (Baskin and Baskin 2004) e.g. such as *C. oxyacantha* and *V. pyramidata*, to come out of dormancy. Pre-chilling at 3°-5°C found to, partly, release dormancy in *C. oxyacantha* (Ellis *et al.*, 1985). Seeds used in the experiment in 2011 were dry stored at ca 5°C for a period of one year. Dry storage at this temperature condition might promote dormancy release. Whereas seeds used in the experiment in 2010 were freshly harvested and presumably in the deeper dormancy status. Ghersa et al., (1992) showed a decreased sensitivity to temperature fluctuation, as a depth sensing mechanism, in after ripened seeds of *Sorghum halepense* that their germinability was increased at constant temperature. They concluded that the decreased sensitivity to fluctuating temperature might be associated with seed aging process.

There are studies reported a similar result (Bliss and Smith, 1985; Grundy et al., 1996 and 2003a). Grundy *et al.*, (2003a) observed that seedling emergence was inhibited in seeds of *Veronica persica*, *V. arvensis*, *Stellaria media*, *C. album*, and *Polygonum aviculare* at soil surface, it was increased in seeds buried at the shallower depths with subsequent decline at the deeper depths. They attributed the observed reduction in emergences in the surface layer to predation or rapid changes in moisture and temperature at the soil surface in their experiment. In the present study, seeds at the soil surface were covered with a fine mesh to avoid predation. Bliss and Smith (1985) showed a two-phase response with germination in *Galiutn aparine* L. and *C. album* reaching a peak at 2 mm depth. They proposed some degree of germination inhibition at the soil surface possibly due to a high rate of cycling between red/far-red photoreceptor phytochrome. High light intensity (25x10-5 mol m-2 s-1) has also been reported to be less suitable for germination than a low light intensity (10-5 mol m-2 s-1) in *Carthamus Lanatus* L. (Ellis *et al.*, 1985). Seed germination in *V. hispanica* was shown to be lower in light and at fluctuating temperature than that of in darkness and at constant temperature (Saatkamp et al., 2011). Moreover, larger-seeded species appear to require more time to imbibe water prior to germination (Forcella et al., 2000). Thus, large seeds at the soil surface are likely to be exposed to greater risk of dehydration than their buried or smaller-seeded counterparts (Buhler, 1995). Consequently, emergence patterns of some species, especially those with larger seeds, may exhibit parabolic relationships with soil burial depth (Forcella et al., 2000) as shown in the present study (Figure 3 and 7).

Soil cultivation measures affect seeds distribution within the soil profile (Grundy *et al.*, 1999; Chauhan and Johnson 2009) and, therefore, subsequent seedling emergence (Buhler and Mester, 1991). For example, low soil disturbance systems (e.g. zero or minimum tillage) may leave a large proportion of weed seeds on or near the soil surface, whereas high soil disturbance systems may buried a large proportion of weed seeds in deeper soil depths (Chauhan and Johnson 2009). Seeds left on the soil surface are in a vulnerable zone because of rapid changes in moisture and temperature and are also more exposed to predation (Grundy *et al.*, 2003a). Thus, cropping practices may be adapted to manage weed seed pool in the cultivations. In addition, buried weed seeds in deeper soil layers can give competitive advantages to the crop over weeds when the seedling emergence of the weeds is delayed by soil depth. Because crop seedlings can emerge and establish earlier than the weeds to use the resources (e.g. water, light and nutrient), and over shade the late emerged weed seedlings.

CONCLUSION

Both species of *C. oxyacantha* and *V. pyramidata* showed the capacity to emerge over relatively extended period, though large numbers of their seedling may emerge in a short period as a main seedling flush. This characteristic is of important value for a weed species to maintain in the cropping systems regardless of weed control operation. Weed control operations may not be practical over a long period of time during a crop development due to economical and practical reasons. Therefore, usually few seedlings may escape weed control practices to sustain future weed infestation. Seedling emergence was low in both species in seeds placed on the soil surface. This could be of adaptation importance for the species to avoid germination, when there is a risk of water shortage at soil surface for seedling survival. Time to start seedling emergence of both species was also delayed by soil depth. This can be considered to adapt soil cultivation tactics to bury a large proportion of the weed seeds in deeper soil depths and subsequently give competitive advantages to the crops over weeds.

REFERENCES

Baskin J.M. and Baskin C.C. (2004) A classification system for seed dormancy. *Seed Science Research* 14: 1-16.

Bassiri A., Rouhani I. and Ghorashy S.R. (1975). Effect of temperature and scarification on germination and emergence of wild safflower, *Carthamus oxyacantha* Bieb.. *The Journal of Agricultural Science* 84: 239-243.

Bassiri A. and Rouhani I. (1976) Effect of seed treatment on germination of wild safflower. *Weed Science* 24: 233-234.

Benvenuti S., MacChia M. and Miele S. (2001) Quantitative analysis of emergence of seedlings from buried weed seeds with increasing soil depth. *Weed Science* 49: (4) 528-535

Bewley J.D. (1997) Seed germination and dormancy. *The Plant Cell* 9: 1055-1066.

Blackshaw R.E. (1992) Soil temperature, moisture, and seed burial depth effects on redstem filaree (*Erodium cicutarium*) emergence. *Weed Science* 40: 204-207.

Bliss D. and Smith H. (1985) Penetration of light into soil and its role in the control of seed germination. *Plant, Cell and Environment* 8: 475–483.

Bouwmeester H.J. and Karssen C.M. (1993) Seasonal periodicity in germination of seeds of *Chenopodium album* L. *Annals of Botany* 72: 463–473.

Buhler D.D. (1995) Influence of tillage systems on weed population dynamics and management in corn and soybean in the central USA. *Crop Science* 35: 1247–1258.

Buhler D.D. and Mester T.C. (1991) Effect of tillage systems on the emergence depth of giant (*Setaria faberi*) and green foxtail (*Setaria viridis*). *Weed Science* 39: 200-203.

Chauhan B.S. and Johnson D.E. (2009) Influence of Tillage systems on weed seedling emergence pattern in rainfed rice. *Soil and Tillage Research* 106: 15-21.

Diggle P.J., Heagerty P., Liang K.Y. and Zeger S..L (2002) *Analysis of Longitudinal Data*. Oxford: Oxford University Press.

Ellis R.H., Hong T.D. and Roberts E.H. (1985) *Handbooks for Genebanks No.2. Handbook of Seed Technology for Genebanks Vol. II. Compendium of Specific Germination Information and Test Recommendations*. International Board for Plant Genetic Resources, Rome, Italy.

Forcella F., Benech-Arnold R.L., Sanchez R. and Ghersa C.M. (2000) Modelling seedling emergence. *Field Crops Research* 67: 123–139.

Ghersa C.M., Benech-Arnold R.L. and Martinez M.A. (1992) The role of fluctuating temperatures in germination and establishment of *Sorghum halepense*. Regulation of germination at increasing depths. *Functional Ecology* 6: 460-468.

Grundy A.C. (2003) Predicting weed emergence: a review of approaches and future challenges. *Weed Research* 43: 1–11.

Grundy A.C., Mead A. and Bond W. (1996) Modelling the effect of weed-seed distribution in the soil profile on seedling emergence. *Weed Research* 36: 375-384.

Grundy A.C., Mead A., and Burston S. (1999) Modelling the effect of cultivation on seed movement with application to the prediction of weed seedling emergence. *Journal of Applied Ecology* 36: 663–678.

Grundy A.C., Mead A. and Burston S. (2003a) Modelling the emergence response of weed seeds to burial depth: interactions with seed density, weight and shape. *Journal of Applied Ecology* 40: 757-770.

Grundy A.C., Peters N.C.B., Rasmussen I.A., Hartmann K.M., Sattin M., Andersson L., Mead A., Murdoch A.J. and Forcella F. (2003b) Emergence of *Chenopodium album* and *Stellaria media* of different origins under different climatic conditions. *Weed Research* 43: 163–176.

Harper J.L. (1977) *Population biology of plants.* Academic Press, London, UK.

Hecker M. and Köhler K.H. (1979) Protein and nucleic acid synthesis during imbibition of dormant and after-ripened *Vaccaria pyramidata* seeds. *Developmental Biology* 69: 270-280.

Hecker M., Köhler K.H., Wachlin G. and Bernhardt D. (1982) Relation between synthesis and germination of *Vaccaria pyramidata* seeds. *Phytochemistry* 21 (7) 1491-1494.

Henson I.E. (1970) The effects of light, potassium nitrate and temperature on the germination of *Chenopodium album*. *Weed Research* 10: 27-39.

Karssen C.M. (1970). The light promoted germination of the seeds of *Chenopodium album* L. III. Effect of the photoperiod during growth and development of the plants on the dormancy of the produced seeds. *Acta Botanica of Netherland* 19: 81-94.

Koller D. (1972) *Environmental control of seed germination.* In: Kozlowski, T.T. (ed.) *Seed Biology*, Vol. II. Academic Press, London, pp. 1–101.

Leblanc M.L., Cloutier D.C. and Hamel C. (2002) Effect of water on common lambsquarters (*Chenopodium album* L.) and barnyardgrass (*Echinochloa crus-galli* (L.) Beauv.) seedling emergence in corn. *Canadian Journal of Plant Science* 82: 855–859.

Leblanc M.L., Cloutier D.C., Stewart K.A. and Hamel C. (2004) Calibration and validation of a common lambsquarters (*Chenopodium album*) seedling emergence model. *Weed Science* 52: (1) 61-66.

Roman E.S., Murphy S.D. and Swanton C.J. (2000) Simulation of *Chenopodium album* emergence. *Weed Science* 48: 217–224.

Saini H.S., Bassi P.K., and Spencer M.S. (1985) Seed germination in Chenopodium album L. relationship between nitrate and the effect of plant hormones. *Plant Physiology* 77: 940-943.

Saatkamp A., Affre L., Dutoit T. and Poschlod P. (2011) Germination traits explain soil seed persistence across species: the case of Mediterranean annual plants in cereal fields. *Annals of Botany* 107(3): 415–426.

Steiner A.M. and Ruckenbauer P. (1995) Germination of 110-year-old cereal and weed seeds, the Vienna Sample of 1877. Verification of effective ultra-dry storage at ambient temperature. *Seed Science Research* 5: 195-199.

Stoller E.W. and Wax L.M. (1973) Temperature variations in the surface layers of an agricultural soil. *Weed Research* 13: 273-282.

Taab A. and Andersson L. (2009a) Primary dormancy and seedling emergence of black nightshade (*Solanum nigrum*) and hairy nightshade (*Solanum physalifolium*). *Weed Science* 57: 526–532.

Taab A. and Andersson L. (2009b) Seed dormancy dynamic and germination characteristics of *Solanum nigrum*. *Weed Research* 49: 490-498.

Tanveer M.Z.F., Tahir M., Javaid M.M. and Khaliq A. (2012) Environmantal factors affecting the germination and seedling emergence of *Carthamus oxyacantha* M. Bieb. (wild safflower). *Pakistan Journal of Weed Science Research* 18(2): 221-235.

Taylorson R.B. (1970) Changes in Dormancy and Viability of Weed Seeds in Soils. *Weed Science* 18 (2): 265-269.

Thompson K. and Grime J.P. (1983) A comparative study of germination responses to diurnally fluctuating temperatures. *Journal of Applied Ecology* 20: 141-156.

Weaver S.E., Tan C.S. and Brian P. (1988) Effect of temperature and soil moisture on time of emergence of tomatoes and four weed species. *Canadian Journal of Plant Science* 68(3): 877-886.

Woolley J.T. and Stoller E.W. *(1978)* Light penetration and light-induced seed germination in soil. *Plant Physiology* 61: 597-600.

In: Weeds and Their Ecological Functions
Editor: Alireza Taab
ISBN: 978-1-62808-332-3

Chapter 3

SEEDLING EMERGENCE OF *CHENOPODIUM ALBUM* IN THE SOIL PROFILE: SUMMER ANNUAL SPECIES

Alireza Taab,* Mehrangiz Akbari and Maryam Mottaghi Shahpar
Department of Agronomy and Plant Breeding,
Faculty of Agriculture, Ilam University, Ilam, Iran

ABSTRACT

Chenopodium album L. is a problematic weed in many crops. Information on seedling emergence characteristics is necessary to optimize the timing of weed control operations. Seedling emergence pattern was studied during two years with seeds of *C. album* buried at soil depths of 0, 2, 4, 6, 8, and 10 cm in outdoor pots. According to the measurements, the mean daily soil temperature was similar at all depths, whereas fluctuation in temperature was higher at the soil surface and decreased with depths. Seedlings emerged from all depths over extended period of time from mid February to mid May. Total numbers of emerged seedlings was higher for seeds placed at the soil surface, possibly due to the stimulatory effect of light and daily fluctuation in temperature on

* Corresponding autor: Alireza Taab, Ph.D., Department of Agronomy and Plant Breeding, Faculty of Agriculture, Ilam University, 69315-516 Ilam, Iran. Tel: (+98) 918 6149226; Fax: (+98) 841 2227015; E-mail: a.taab@mail.ilam.ac.ir.

germination, and decreased with depth. However, seedling emergence surprisingly increased again in seeds buried at the depth of 10 cm with a main flush in late March and few sporadic emergences before and after this time. This information can be used to predict seedling emergence behavior in different soil depths and therefore use in management of *C. album* in agriculture.

1. INTRODUCTION

Chenopodium album L. (common lambsquarters) is one of the world's worst weeds found in many crops throughout the world (Holm et al., 1977). It can grow well under a wide range of climate and soil conditions (Mitich, 1988). It has been reported to cause a crop yield reduction of 36% in barley (*Hordeum vulgare* L.), (Conn and Thomas, 1987), 11% in corn (*Zea mays* L.) (Beckett et al., 1988), 48% in sugarbeets (*Beta vulgaris* L.), (Schweizer, 1983), and 36% in tomato (*Lycopersicon esculentum* Mill.) (Bhowmik and Reddy, 1988) and up to 65% in lettuce (*Lactuca sativa* L.) (Santos et al., 2004). Therefore, the control of this species is of high importance.

Information on seedling emergence characteristics of a weed species helps to improve prediction of emergence timing of the species in a field condition (Taab and Andersson, 2009a). Data on emergence ability in response to burial depths is crucial to develop predictive tools (Forcella et al., 2000) to optimize the timing of weed control measures (Grundy, 2003). Soil disturbances can be adapted to manage weed seed banks. For instance, deeply buried seeds may not be able to emerge or zero tillage may promote greater emergence of species that require light for germination (Chauhan and Johnson, 2009).

Emergence timing of a weed species may be affected by several factors. For example, seed dormancy and germination characteristics, soil temperature, moisture and depths of burial in the soil. Variation in seed dormancy and morphology of *C. album* has been reported by Karssen (1970). Germination characteristic of this species has also been well studied. For instance, Boumeester and Karssen (1993) found seasonal periodicity in seed dormancy of *C. album* in the field. Light and nitrate promoted germination of this species (Boumeester and Karssen, 1993). Saini (1985) reported germination of *C. album* at the range of 12 to 32C. A water potential of –640 kPa was required for germination of *C. album* to occur (Roman et al., 2000). Seedling emergence of *C. album* has also been studies across different climatic conditions (Grundy et al., 2003b).

Emergence studies were conducted by recording seedlings of *C. album* emerged from natural weed infestation (Leblanc et al., 2004) or by sowing seeds at specific soil depths (Weaver et al., 1988). The number of seedlings and rate of seedling emergence was shown to decrease with increasing depth of burial in many species (Benvenuti et al., 2001). Grundy et al., (1996 and 2003a) found that total seedling emergence of *C. album* substantially declined with increasing depth of burial. Also emergence was inhibited at the soil surface. However, there is no report on the effect of burial depth on emergence pattern of the species.

Environmental factor required for seed germination (i.e. temperature, moisture and light) may vary within the soil profile (Harper, 1977; Stoller and Wax, 1973). The variation in these environmental factors was found to affect fate of seeds in the soil (Taylorson, 1970; Stoller and Wax, 1974). Less than 1% of light penetrates only few millimeters into soil (Woolley and Stoller, 1978). Temperature was found to be the major factor affecting weed seedling emergence pattern (Henson, 1970; Blackshaw, 1992; Leblanc et al., 2002).

There is an increased tendency toward optimizing the timing of weed control measures and therefore study is needed to understand factors controlling the time and extent of seedling emergence in the field. The objective of this study was to determine the effect of seed burial depths on seedling emergence pattern of *C. album* in a field condition.

2. Material and Methods

2.1. Seed Collection

Seeds were collected in early January 2010 for the experiments conducted in 2010 and 2011 from plants growing in a research station (33° 38' 17" North, 46° 25' 50" East) in Ilam, Iran. The seeds were cleaned and kept at ca 5°C until use.

2.2. Sowing Seeds in Pots Outdoors

In mid January 2010 numbers of pots (18-cm diameter, 18-cm height) were prepared for sowing the seeds. The bottom of the pots was open with several holes to ensure similar moisture conditions to the outside pots. The pots were filled and packed with clay soil (clay: 42%; silt: 24% and sand 34%,

PH: 7.85 and organic carbon: 1.16%) to nominated depths in each pots so that the surface was measured from 5 cm below the rim. The soil used was extracted from the deep layer of soil (>30 cm; to ensure lack of weed seed infestation) from an experimental field and it was sieved and bulked under plastic film cover for 4 months. The pots were then arranged in a row; each row represented one block, and sunk into the ground with their rims protruding 5 cm above the ground level.

100 seeds were evenly distributed in each pot at 6 soil depths of 0, 2, 4, 6, 8, and 10 cm. The pots were then filled with the soil to 5 cm below the rim. The pots with seeds placed at soil surface were covered with a fine mesh to prevent any possible seed predation. The experimental design was *split-block* (*strip-plot*) consisting of six treatments (seed burial depths) with four replicates. The experiment was repeated in mid January (12/13th) 2011 using a clay soil (clay: 40%; silt: 26% and sand 34%, PH: 7.60 and organic carbon: 3.6%) as mentioned above.

2.3. Soil Temperature

Soil temperature was recorded at all depths during the experiment using temperature loggers (iButton 1-Wire Thermochron DS1921G-F5, Dallas Semiconductor, Inc. US).

2.4. Seedling Emergence

The pots were surveyed for emergence everyday. The emerged seedlings were counted and removed until the end of the season when no more seedlings were observed. The pots were watered thoroughly when needed using a hand sprinkler during the experiment. The care was taken to avoid any soil disturbances.

2.5. Statistical Analysis

The analysis was conducted to test the differences between years, burial depths, and interaction between these factors using the GENMOD procedure (SAS Institute Inc. 2001). The proportion of emerged seedlings was tested as a categorical variable, assuming binomial distribution and logit link function.

We used a robust method (GEE) to compensate for overdispersion (Diggle *et al.* 2002). Standard error of mean (SEM) was calculated using the LSmeans statement with pdiff option in the GENMOD procedure with account for the assumed binomial distribution. SEMs were then recalculated using back transformation of estimates from the logit scale to the original scale by applying the inverse link function.

3. Results

In 2010, seeds placed at soil surface started to emerge sporadically in February and this continued until late May. For seeds buried at the depths of 2, 4, 6 and 8 cm few seedlings were emerged. There was main seedling flush in late March for seeds buried at the depth of 10 cm (Figure 1). In 2011, seeds placed at soil surface started to emerge in mid February and continue until early May with a larger flush in early April. This extended emergence pattern was similar for seeds buried at the depths of 2, 4, and 6 cm. There was one main seedling flush for seeds buried at the depths of 8 and 10 cm in late March with quite few seedlings emerged before and after this period. This main flush of seedlings was also shown at the same time, but to a less extent, in seeds buried at the depth of 4 and 6 cm (Figure 2). In both years, a high fluctuation in temperature was observed at the soil surface. The mean daily soil temperature was similar at all soil depths, whereas fluctuation in temperature was decreased with increasing depths. There were significant differences between years and among sowing depths in total emerged seedlings (Table 1). Total seedling emergence was high for seeds placed at the soil surface in 2010 that decreased in seeds buried in the soil depths of 2 to 8 cm. Surprisingly, total seedling emergence was increased again in seeds buried at the depth of 10 cm. A similar pattern was also observed in 2011 but with a larger number of emerged seedlings (Figure 3).

Table 1. Analysis of deviance for seedling emergence of *Chenopodium album* from soil depths in 2010 and 2011

Source	DF	Chi-Square	P-value*
Year	1	24.91	<.0001
Sowing Depth	5	15.44	0.0087
Year* Depth	5	7.22	0.2048

*P< 0.05 is used as criterion for significance.

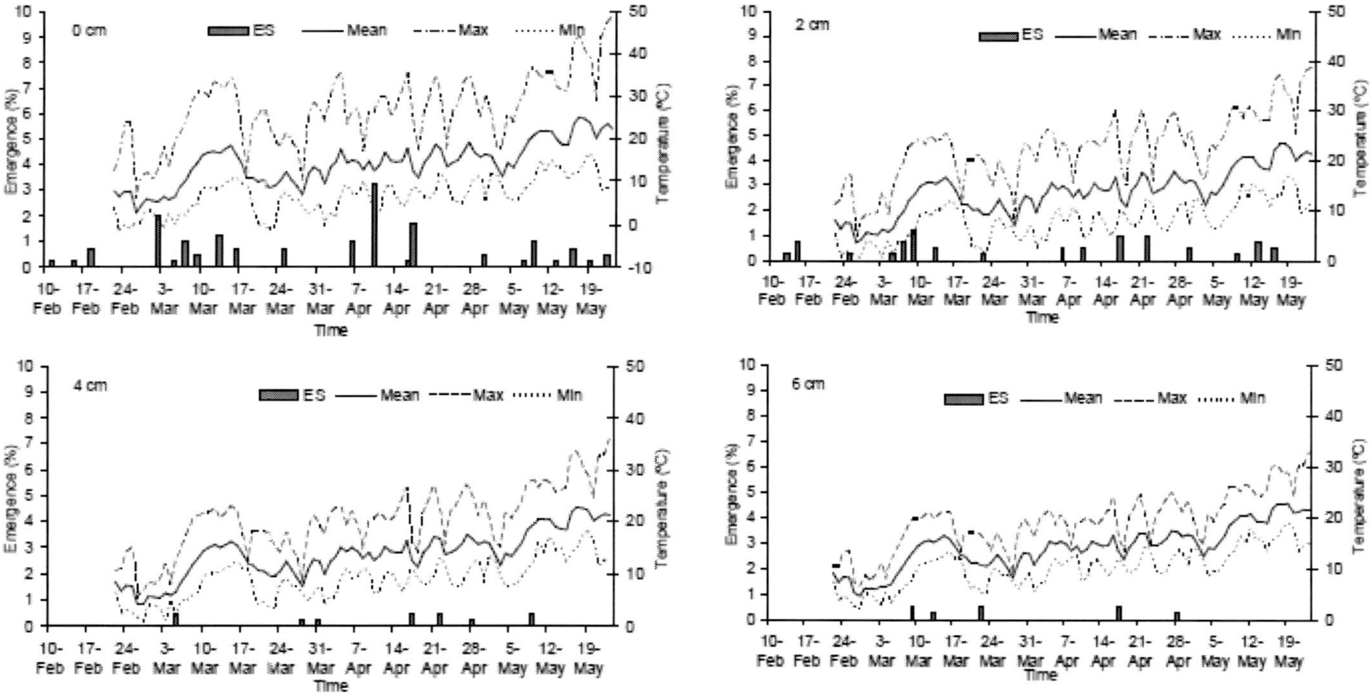

0 cm
2 cm
4 cm
6 cm
ES
Mean
Max
Min
Emergence (%)
Temperature (°C)
Time
10- Feb
17- Feb
24- Feb
3- Mar
10- Mar
17- Mar
24- Mar
31- Mar
7- Apr
14- Apr
21- Apr
28- Apr
5- May
12- May
19- May

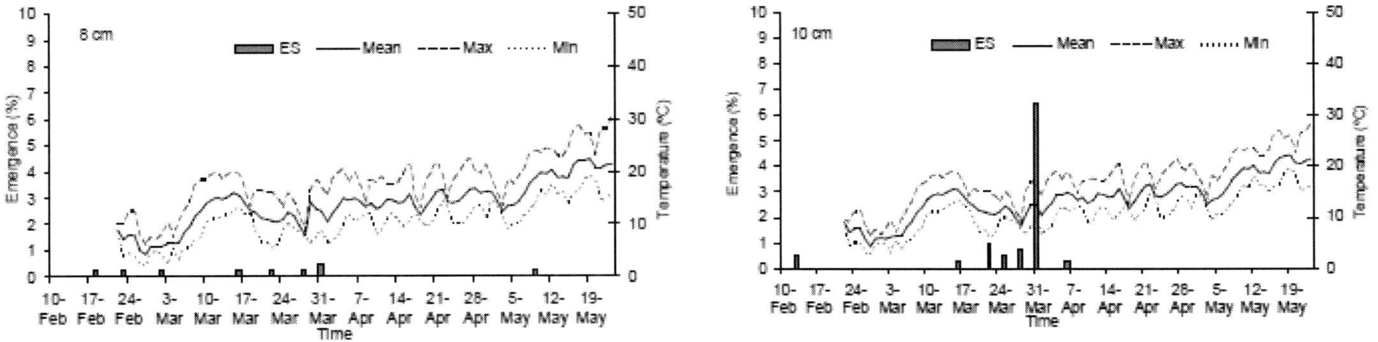

Figure 1. Seedling emergence pattern of *Chenopodium album* from burial depths of 0 (seeds sown on the soil surface), 2, 4, 6, 8, and 10 cm. ES: emerged seedlings; Mean, Max, and Min: mean, maximum, and minimum soil temperatures, respectively, measured at the burial depths in the pots outdoors in 2010.

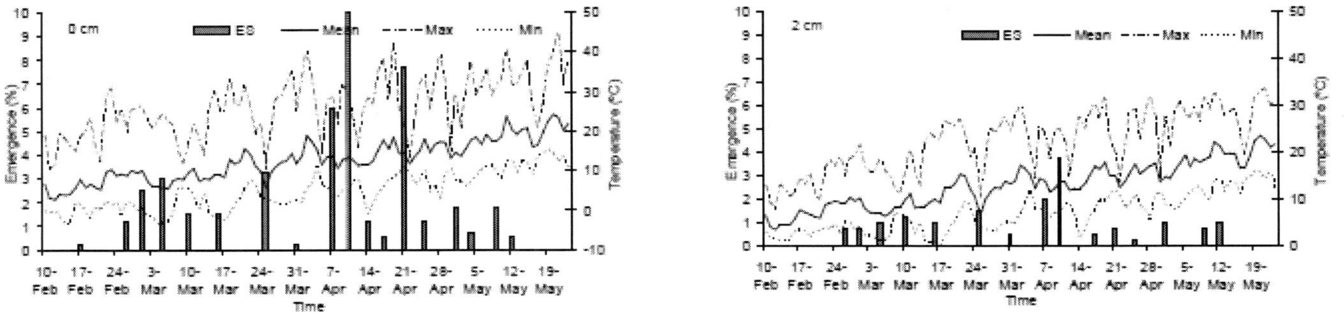

Figure 2. (Cotinued)

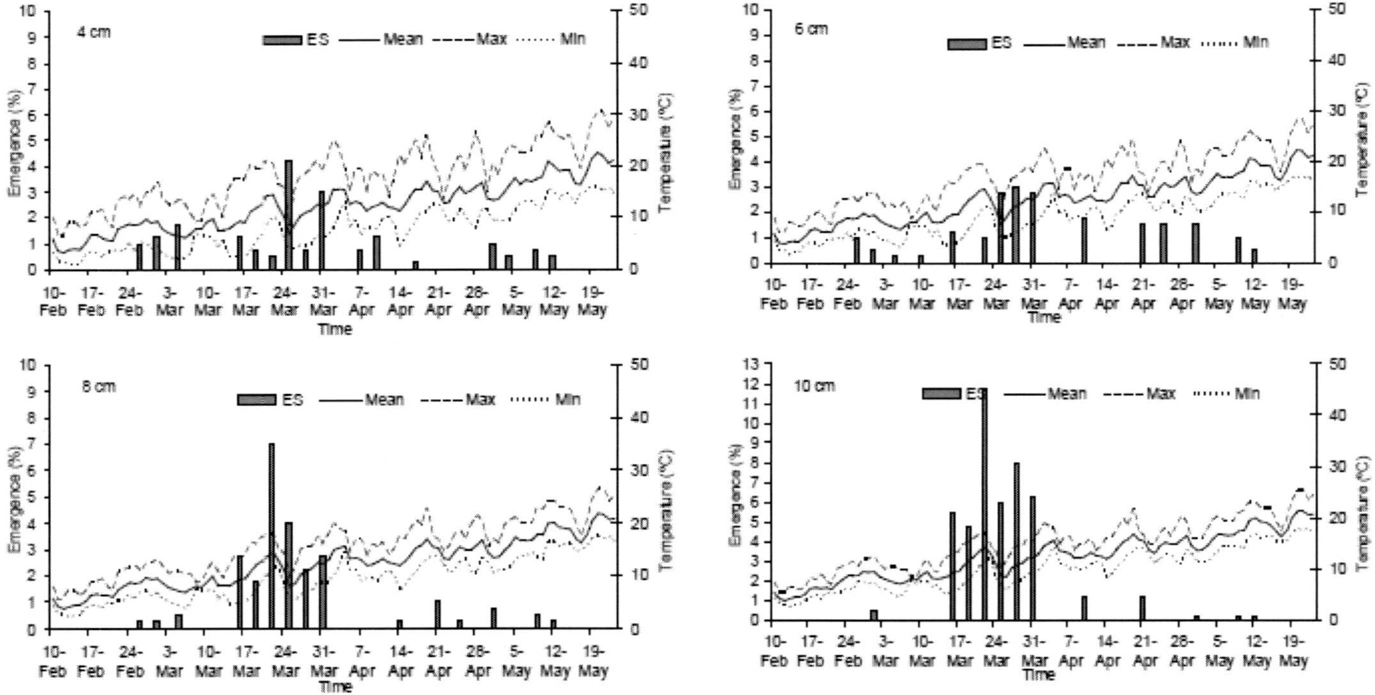

Figure 2. Seedling emergence pattern of *Chenopodium album* from burial depths of 0 (seeds sown on the soil surface), 2, 4, 6, 8, and 10 cm. ES: emerged seedlings; Mean, Max, and Min: mean, maximum, and minimum soil temperatures, respectively, measured at the burial depths in the pots outdoors in 2011.

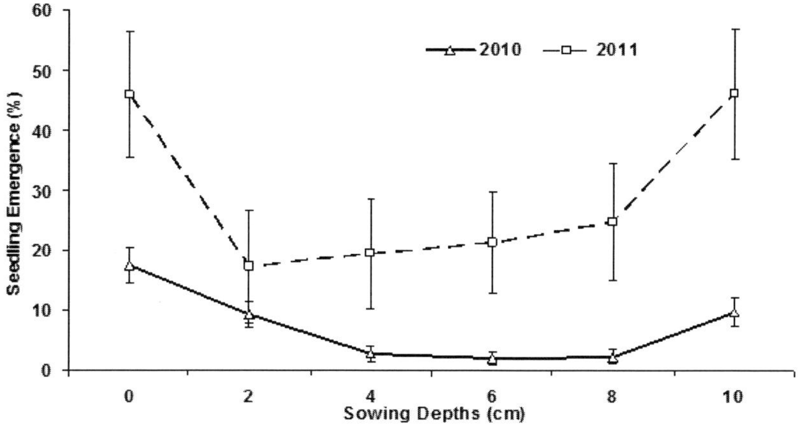

Figure 3. Total number of seedling of *Chenopodium album* emerging from different soil depths. Vertical bars represent SEM.

4. Discussion

Seedlings of *C. album* showed a capacity to emerge over an extended period from almost all burial depths, although the numbers of emerged seedlings varied with depths. Seedling emergence over an extended period may contribute the seriousness of a weed species in agricultural conditions. This characteristic at least enables portions of the emerged seedlings to escape weed control operations or climatic extremes (Taab and Andersson, 2009a). This could also be a reason why *C. album* is one of the world's worst weeds (Holm *et al.*, 1977) regardless of long time effort for control it in agriculture. Bi- or three-modal seedling emergence pattern over an extended period has been shown in many weed species (Taab and Andersson, 2009a; Myers et al., 2005; Ogg and Dawson, 1984; Stoller and Wax, 1973). This characteristic is of important adaptation value for a weed species to survive in agricultural areas. Seedling started to emerge in February when the daily mean soil temperature increased above ca 5°C (Figure 1and2). The emergence was continued sporadically by increasing of daily temperature until mid May in both years when the maximum daily temperature rose above 25°C. This is in agreement with report of germination for *C. album* in the field at temperatures ranging from 5 to 25°C by Bouwmeester and Karssen (1993). They also reported release of dormancy in buried seeds *of C. album* during winter and spring.

For *C. album*, release of seed dormancy due to stratification at 3°C or buried in soil outdoor has been reported (Milberg and Andersson, 1998). Williams and Harper (1965) recognized four categories of polymorphic seeds in *C. album* differing in dormancy and germination requirements. They found out some seeds readily germinate without pre-chilling but some seeds need a period of chilling (e.g. at 5°C) before germination to take place. A similar seed polymorphism response in *C. album* has also been reported by Henson (1970). In the present study, it is likely that non-dormant seeds or seeds with a lower level of dormancy germinated in February and seeds with stronger level of dormancy overcome their dormancy later in the season due to the temperature conditions during the experiment. With the passage of time, more seeds relived their dormancy and germinated when faced with the range of temperature appropriate for germination. This characteristic enables the species to extend its seedling emergence over an extended period.

There are reports on seedlings emergence of *C. album* over a relatively extended period, however it is unclear how the pattern is affected by soil depths. For instances, seedlings of *C. album* began to emerge from middle to late March and emerged with peak of emergence from the end of March until the end of April. Thereafter, the seedling continued to emerge in low numbers until September (Ogg and Dawson, 1984). Håkansson (1983) reported seedling emergence of *C. album* from March to August with a peak in May. Williams (1963) also reported seedling emergence of *C. album* throughout the summer with two peaks in emergence, one in late April and another in August. Leblanc et al., (2002) also reported emergence of *C. album* from May to July with a peak in late May. From the present study, it is concluded that there are three responses in terms of total emerged seedlings and emergence timing pattern. i) Large numbers of seedlings may emerge at the topsoil over an extended period. ii) Total numbers of emerged seedling may decrease with depth at the medium depths (i.e. 2-6 cm). iii) Seeds buried at the deeper soil layers showed one main flush of seedlings emergence with an increased number of seedlings in late March (Figures 1 - 3).

Seedling emergence was stopped in mid May in both years when daily mean soil temperature raised above 20°C with the maximum soil temperature above 25°C. It might be that this soil temperature regime during this period caused induction of secondary dormancy. Bouwmeester and Karssen (1993) observed start of dormancy induction in the field when temperature increased to ca 13 to 20°C. Control of germination and emergence timing by dormancy induction is of ecological importance to subsequent survival of seedlings and reproduction (Bewley, 1997).

Seedlings that emerge too late may face to sever competition with crop stands. Moreover, they may die with onset of frost in autumn before reproduction.

Total seedling emergence counts were higher for seeds at the soil surface which declined with increasing burial depths with an increase at the depth of 10 cm in both years (Figure 3). An increased number of emerged seedling at the soil surface could be due to the stimulatory effect of light and greater fluctuation in temperature on germination. Alternating temperature and light were found to promote germination in seeds of *C. album* (Henson, 1970; Saini et al., 1985). Seeds buried in the soil did not receive light and fluctuation in temperature was also decreased with depth until it became similar to constant temperature condition (Figure 1 and 2) which might be less favored to stimulate germination. Taab and Andersson (2009b) studying the effect of alternating temperature on germination showed optimal amplitude above and below which germination decreased.

However, there are studies reported a different result (Bliss and Smith, 1985; Grundy et al., 1996 and 2003*a)*. Bliss and Smith (1985) showed a two-phase response with germination in *C. album* reaching a peak at 2 mm depth. They proposed some degree of germination inhibition at the soil surface possibly due to a high rate of cycling between red/far-red photoreceptor phytochrome. Grundy et al., (1996 and 2003a) observed that seedling emergence was low in seeds of *C. album* at soil surface, it increased in seeds buried at the shallower depth with subsequent decline at the deeper depths. They attributed the observed reduction in emergence in the surface layer to predation or rapid changes in moisture and temperature at the soil surface in their experiment. In the present study seeds at the soil surface were covered with a fine mesh to avoid predation. As concluded above increased number of seedling emergence for seeds at soil surface might be due to stimulatory effect of fluctuating temperatures and light on seed germination of the species.

Surprisingly, total seedling emergence was increased at the depth of 10 cm in both years (Figure 3). A similar pattern was also observed in *Amaranthus retroflexus*, *A. blitoides*, *Brassica kaber* and *Solanum nigrm* (unpublished data). To our knowledge this phenomenon has not been observed before. This remains to be thoroughly examined experimentally. It could be examined based on the following possible speculations. i) After rainfall or irrigation, the water present in the upper soil layers will move downward due to drainage or evaporated from the topsoil into the atmosphere (Brouwer et al., 1985). Even though large numbers of seeds germinated at soil surface due to the effect of light and fluctuation in temperature, it is more likely that upper layers of soil experienced fluctuation in soil water content than that of deeper layers e.g. 10 cm. Therefore, seeds in the upper layers might have been exposed to such a temporary fluctuation in soil moisture content between irrigation or rainfall events while seeds buried at the depth of 10 cm might have been exposed to stable soil moisture content. ii) It could also be due to the effect of local adaptation. Seed populations from dryer areas like Iran (site of collected seeds) need to ensure that the amount of moisture required for early seedling growth and establishment would be available before emergence to take place. Insufficient soil moisture is known as a factor to cause pre-emergence mortality (Grundy *et al.*, 2003a). Shallower soil layers usually experience high fluctuations in moisture that is detrimental for seedlings survival. Seedlings would emerge from deeper soil depths predicting availability of sufficient moisture for seedling establishment.

Total seedling emergence was higher significantly (Table 1) in 2011 than in 2010 (Figure 3). This might be due to the following reasons. i) Fresh seeds soon after collection were used in experiment in 2010. Large proportions of fresh seeds of *C. album* might be dormant as shown by Williams and Harper (1965), Henson (1970), and Bouwmeester and Karssen (1993). Dormant seeds may need a period of time conditioning at proper temperature regimes to come out of dormancy and subsequently germinate (Bouwmeester and Karssen 1993). The temperature conditions and time required for dormant seeds to come out of dormancy might have not been fulfilled in 2010. Therefore, seedling emergence was poor in 2010. ii) Seeds used in experiment in 2011 were dry stored at ca 5°C for a period of one year. Dry storage at this temperature condition might promote dormancy release. iii) In addition, temperature during winter in 2011 was slightly lower than in 2010. In *C. album*, Grundy et al. (2003b) stated that deeper winter chilling may have a greater dormancy releasing effect than a mild winter chill, hence releasing a greater proportion of the population to be available for germination. Similarly, Leblanc et al., (2002) observed larger numbers of emerged seedling in a year with cooler soil temperature than warmer one. Thus, it is likely that more seeds came out of dormancy and germinated in 2011.

Soil cultivation measures affect seed distribution within the soil profile (Grundy et al., 1999; Chauhan and Johnson 2009) and subsequent seedling emergence (Buhler and Mester, 1991). For example, low soil disturbance systems (e.g. zero or minimum tillage) may leave a large proportion of the weed seeds on or near the soil surface, whereas high soil disturbance systems may buried a large proportion of weed seeds in deeper soil depths (Chauhan and Johnson 2009).

Seeds left on the soil surface are in a vulnerable zone because of rapid changes in moisture and temperature and are also more exposed to predation (Grundy et al., 2003a). Cropping practices with minimum soil disturbances may be adapted to rapid seed pool depletion either by exposure of seeds to environmental hazards or stimulation of germination due to effect of temperature alteration and light on soil surface.

The practices that bury seeds below the soil surface may allow them to persist, e.g. seeds buried in the medium depths. Portions of seeds buried deep in the soil (e.g. 10 cm) may yield an increased number of seedling emergence in one main flush that can be targeted by a single weed control measure.

CONCLUSION

In conclusion, seeds of *C. album* have the ability to germinate over an extended period in the field especially where they are placed on topsoil or in shallower depths. This characteristic makes the species very difficult to control.

Because weed control operation over a long period of time is costly, in addition, it is usually not allowed during a season due to the crops developmental characteristics.

For instance, herbicide application may not be possible during most of the crop developmental stages or repeated mechanical weed control operation during advanced crop development may also cause sever damage to the crop. Seedling emergence of *C. album* was not decreased by increasing of burial depths as expected (e.g. Grundy et al., 2003a).

Therefore, the emergence response of this species to burial depth is not only species-specific (Mohler, 1993) but also could be population-specific. The present study provides the information to predict the emergence behavior of *C. album* in the cultivations and further can be used in integrated weed management systems.

Seedlings of this species emerge early in spring, therefore, some weed control is likely possible through seedbed preparation. For control of the later emerging seedlings, mechanical weed control can be applied.

In addition, to efficient control of this species in the cultivations, (i) cultural methods that enable a crop to build up its canopy early in the season and (ii) crops with high competitive ability need to be considered in crop sequences.

REFERENCES

Beckett, T. H., Stoller, E. H. and Wax, L. M. (1988) Interference of four annual weeds in corn (*Zea mays*). *Weed Science* 36: 764–769.

Benvenuti, S., MacChia, M. and Miele, S. (2001) Quantitative analysis of emergence of seedlings from buried weed seeds with increasing soil depth. Weed Science 49: (4) 528-535

Bewley, J. D. (1997) Seed germination and dormancy. *The Plant Cell* 9: 1055-1066.

Bhowmik, P. C. and Reddy, K. N. (1988) Interference of common lambsquarters (*Chenopodium album*) in transplanted tomato (*Lycopersicon esculentum*). *Weed Technology* 2: 505–508.

Blackshaw, R. E. (1992) Soil temperature, moisture, and seed burial depth effects on redstem filaree (*Erodium cicutarium*) emergence. *Weed Science* 40: 204-207.

Bouwmeester, H. J. and Karssen, C. M. (1993) Seasonal periodicity in germination of seeds of *Chenopodium album* L. *Annals of Botany* 72, 463–473.

Brouwer, C., Goffeau, A. and Heibloem, M. (1985) *Irrigation Water Management: Training Manual No. 1 - Introduction to Irrigation*. Food and Agriculture Organzation of the United Nations (FAO).

Buhler, D. D. and Mester, T. C. (1991) Effect of tillage systems on the emergence depth of giant (*Setaria faberi*) and green foxtail (*Setaria viridis*). *Weed Science* 39: 200-203.

Chauhan, B. S. and Johnson, D. E. (2009) Influence of Tillage systems on weed seedling emergence pattern in rainfed rice. *Soil and Tillage Research* 106: 15-21.

Conn, J. S. and Thomas, D. L. (1987) Common lambsquarters (*Chenopodium album*) interference in spring barley. *Weed Technology* 1: 312–313.

Diggle, P. J., Heagerty, P., Liang, K. Y., and Zeger, S. L. (2002) *Analysis of Longitudinal Data*. Oxford: Oxford University Press.

Forcella, F., Benech-Arnold, R. L., Sanchez, R., and Ghersa, C. M. (2000) Modelling seedling emergence. *Field Crops Research* 67: 123–139.

Grundy, A. C. (2003) Predicting weed emergence: a review of approaches and future challenges. *Weed Research* 43: 1–11.

Grundy, A. C., Mead, A. and Bond, W. (1996) Modelling the effect of weed-seed distribution in the soil profile on seedling emergence. *Weed Research* 36: 375-384.

Grundy, A. C., Mead, A. and Burston, S. (2003a) Modelling the emergence response of weed seeds to burial depth: interactions with seed density, weight and shape. *Journal of Applied Ecology* 40: 757-770.

Grundy, A. C., Peters, N. C. B., Rasmussen, I. A., Hartmann, K. M., Sattin, M., Andersson, L., Mead, A., Murdoch, A. J., and Forcella, F. (2003b) Emergence of *Chenopodium album* and *Stellaria media* of different origins under different climatic conditions. *Weed Research* 43: 163–176.

Harper, J. L. (1977) *Population biology of plants*. Academic Press, London, UK.

Håkansson, S. (1983) Seasonal variation in the emergence of annual weeds – an introductory investigation in Sweden. *Weed Research* 23: 313–324.

Henson, I. E. (1970) The effects of light' potassium nitrate and temperature on the germination of *Chenopodium album*. *Weed Research* 10: 27-39.

Holm, L. G., Plunkett, D. L., Pancho, J. V., and Herberger, J. P. (1977) *The World's Worst Weeds - distribution and biology*, University Press of Hawaii.

Karssen, C. M. (1970). The light promoted germination of the seeds of *Chenopodium album* L. III. Effect of the photoperiod during growth and development of the plants on the dormancy of the produced seeds. *Acta Botanica of Netherland* 19: 81-94.

Leblanc, M. L., Cloutier, D. C. and Hamel, C. (2002) Effect of water on common lambsquarters (*Chenopodium album* L.) and barnyardgrass (*Echinochloa crus-galli* (L.) Beauv.) seedling emergence in corn. *Canadian Journal of Plant Science* 82: 855–859.

Leblanc, M. L., Cloutier, D. C., Stewart, K. A., and Hamel, C. (2004) Calibration and validation of a common lambsquarters (*Chenopodium album*) seedling emergence model. Weed Science 52: (1) 61-66.

Milberg, P. and Andersson, L. (1998) Does cold stratification level out differences in seed germinability between populations? *Plant Ecology* 134: 225–234.

Mitich, L. W. (1988) Common lambsquarters. *Weed Technology* 2: 550–552.

Mohler, C. L. (1993) A model of the effect of tillage on emergence of weed seedlings. *Ecological Applications* 3: 53-73.

Roman, E. S., Murphy, S. D. and Swanton, C. J. (2000) Simulation of *Chenopodium album* emergence. *Weed Science* 48: 217–224.

Saini, H. S., Bassi, P. K. and Spencer, M. S. (1985) Seed germination in *Chenopodium album* L. relationship between nitrate and the effect of plant hormones. *Plant Physiology* 77: 940-943.

Santos, B. M., Duskyb, J. A., Stallb, W. M., and Gilreatha, J. P. (2004) Influence of common lambsquarters (*Chenopodium album*) densities and phosphorus fertilization on lettuce. *Crop Protection* 23: 173–176.

Schweizer, E. E. (1983) Common lambsquarters (*Chenopodium album*) interference in sugarbeets (*Beta vulgaris*). *Weed Science* 31: 5–7.

Stoller, E. W. and Wax, L. M. (1973) Temperature variations in the surface layers of an agricultural soil. *Weed Research* 13: 273-282.

Stoller, E. W. and Wax, L. M. (1974) Dormancy changes and fate of some annual weed seeds in the soil. *Weed Science* 22: (2) 151-155.

Taab, A. and Andersson, L. (2009a) Primary dormancy and seedling emergence of black nightshade (*Solanum nigrum*) and hairy Nightshade (*Solanum physalifolium*). *Weed Science* 57: 526–532.

Taab, A. and Andersson, L. (2009b) Seed dormancy dynamic and germination characteristics of *Solanum nigrum*. *Weed Research* 49: 490-498.

Taylorson, R. B. (1970) Changes in Dormancy and Viability of Weed Seeds in Soils. *Weed Science* 18: (2) 265-269.

Taylorson, R. B. and Borthwick, H. A. (1969) Light infiltration by foliar canopies; significance for light controlled weed seed germination. *Weed Science* 17: 144-147.

Weaver, S. E., Tan, C. S. and Brian, P. (1988) Effect of temperature and soil moisture on time of emergence of tomatoes and four weed species. *Canadian Journal of Plant Science* 68: (3) 877-886.

Williams, J. T. (1963) Biological flora of the British Isles. *Chenopodium album* L. *Journal of Ecology* 51: 711-725.

Woolley, J. T. and Stoller, E. W. (1978) Light penetration and light-induced seed germination in soil. *Plant Physiology* 61: 597-600.

In: Weeds and Their Ecological Functions
Editor: Alireza Taab
ISBN: 978-1-62808-332-3

Chapter 4

WEEDS IN PERENNIAL CROPS AS AN UNEXPECTED TOOL OF INTEGRATED CROP MANAGEMENT

Ilias S. Travlos*

Lecturer of Agronomy, Laboratory of Agronomy, Faculty of Crop Science, Agricultural University of Athens, Athens, Greece

ABSTRACT

The presence of weeds in perennial crops (orchards, olives, vineyards) is certainly not always harmful.

Obviously, this effect depends on the kind of the present weeds and their relative abundance. Indeed, some weeds (especially those characterized as perennial or invasive) are highly competitive while they could cause several other problems to the crops.

However, the presence of some weeds, especially members of Fabaceae or Poaceae and some other exceptional species (e.g. *Oxalis pes-caprae* L.) should be clearly reconsidered regarding their utility as cover species. Some of these weeds have beneficial impacts by causing soil enrichment and stabilization of semi-stable areas and orchards or being valuable forage.

* Corresponding author: Ilias. S. Travlos, Ph.D. Laboratory of Agronomy, Faculty of Crop Science, Agricultural University of Athens, 75, Iera Odos st., 11855 Athens, Greece. Tel: 00302105294483; Fax: 00302105294482; E-mail: htravlos@yahoo.gr.

Moreover, in addition to competition-based weed suppression; certain legumes used as cover crops (hairy vetch, red clover, sweetclover etc.) are also known to suppress weeds through their allelopathic activity both during growth and after their death. Several studies have been conducted in order to evaluate the efficacy of different management systems involving the use of natural vegetation, with respect to the tillage systems generally used in semiarid environments. Future research should focus on legumes or other herbaceous species which can suppress weeds, won't compete for water and nutrients, increase N and organic matter content and control erosion. Such an approach could certainly maintain and enhance the profitability of these sustainable agriculture systems, especially in the case of organic or low input agriculture.

1. INTRODUCTION

In the past decades, agriculture primarily and unwarily depended on the use of herbicides to suppress weeds and ensure high yields. Under that approach, the traits important for weed competition were not among the major goals for breeders, because herbicides seemed to adequately control weeds in practical farming.

Therefore, several alternatives towards weed management have been considered very little in total in crop breeding programs and scientific researches. The application of weed and pest controlling chemical agents has therefore steadily increased, although a number of pesticides have had well-documented negative consequences on the environment and on human health, while several other problems (e.g. herbicide resistance) progressively emerged (Owen and Powles, 2009; Travlos and Chachalis, 2010; Travlos et al., 2010).

Nowadays, weed management in agriculture aimed at the reduction of herbicide use needs to comprise multiple strategies, among them the use of plant species with improved competitive traits or enhanced allelopathic activity (Jordan, 1993; Christensen, 1995; Didon and Hansson, 2002, Travlos et al., 2008). Interest in integrated approaches to crop and pest (including weed) management is being driven not only by declining crop prices coupled with increased input costs but also by consumer concerns about the environment, and health effects of pesticides. In particular, Integrated Weed Management (IWM) essentially means the integration of several practices, including herbicides, to reduce the negative impact of weeds on crops (Anderson, 2007).

Among these practices, the purposed encouragement of specific and less competitive plant species seems crucial towards the direction of a more effective crop management especially in the case of perennial crops. In fact, even if total ground cover control eliminates competition for water and nutrients, it does, however, have many disadvantages regarding soil erosion, frost damage and loss of organic matter. For these and for many more reasons that are going to be discussed, total vegetation control is not environmentally sustainable in our days.

2. Worst-Case Scenarios Related to the Presence or Absence of Weeds

The negative effects of weeds in orchards are already well known and adequately studied (Travlos et al., 2011). Weeds compete with cultivated plants for light, water and nutrients. Weed competition in early summer can significantly decrease shoot growth, tree vigour and fruit bud formation. Weeds often have an advantage because they usually have denser and better developed root systems than young orchard plants and they absorb water and mineral nutrients faster and more than many crops.

Moreover, uncontrolled natural vegetation around young trees is an excellent place for tree-damaging rodents to spend the winter (Lipecki, 2006). Weeds can also provide some orchard pests with good conditions for development and they make the mechanical lifting of young trees in the nursery much more difficult, and thus more expensive (Lipecki, 2006). Allelopathic effects between weeds and cultivated plants are also common (Travlos and Paspatis, 2008). Indeed, in many cases it has been found that competitive effects are caused not only by antagonism for nutrients but in part by phytotoxic substances produced by the weeds. Horowitz (1973) has extensively studied the severe competitive effects of *Sorghum halepense*, *Cyperus rotundus* and *Cynodon dactylon* on one-year-old sour orange seedlings *(Citrus aurantium)* and highlighted the need of an integrated weed management approach especially for young trees.

Weed competition for light is not often a major problem with established trees. However, large weeds in newly planted orchards can shade and severely stunt young trees. Vining plants like *Smilax* spp. or *Toxicodendron radians* can climb mature trees and compete for light, especially in warm climates and semi-abandoned or old orchards.

Similar and intense competition is caused by many other tall plants, common in many orchards of Mediterranean countries and elsewhere such as *Conyza canadensis*, *S. halepense* and *Malva sylvestris*. In that case it could be said that the primary competition between weeds and trees is for water, especially with such vigorous weeds, young trees and during the driest periods of the year.

Water is essential for maximizing fruit size and especially in dry months and years weed populations should be clearly controlled and eliminated. Moreover, the weeds can be easily propagated by both vegetative organs and seed and for this reason weeds such as *S. halepense, C. dactylon* and *Convolvulus arvensis* are extensively dispersed. The vegetative parts of these plants can also survive and regenerate new plants, even after having been buried in the soil.

In addition, many weeds produce enormous amounts of easily dispersed seeds. For example, *Conyza* spp. are prolific seed producers, with a single plant capable of producing thousands of non-dormant seeds (Weaver, 2001), which can be widely dispersed by wind likewise many members of the Asteraceae (Shields et al., 2006). *Conyza* spp. is a notable case since it has progressively become common and one of the most problematic weeds in agronomic crops (Weaver, 2001) because of its adaptation to periodically plant-free, undisturbed soil and establishment in the absence of tillage (Brown and Whitwell, 1988; Buhler, 1992).

Therefore, the opportunistic nature of *Conyza* in undisturbed areas makes them well suited for becoming established in orchards and vineyards, especially in no-tillage or minimum-tillage crop production systems (Brown and Whitwell 1988; Bruce and Kells 1990). Taking into account that over the last two decades the percentage of crop hectares in conservation-tillage has significantly increased (Young, 2006), weeds such as *Conyza* spp. are becoming problems of great concern (Travlos and Chachalis, 2010).

Many annual and perennial weeds are found in citrus and olive groves and other orchards. Traditionally, orchards are mechanically cultivated or hoed by hand to remove weeds. These are labor-intensive and time-consuming activities, while they favour soil erosion and loss of organic matter (Shui et al., 2004). Effective use of herbicides can certainly solve some of these problems. Glyphosate and many broad-spectrum and residual herbicides have been and still are used with more or less succession (Baylis, 2000). Fruit growers strongly believe that all these negative effects that were previously mentioned may dramatically reduce yields, and have often attempted to eliminate all weeds from their orchards, mainly by using broad-spectrum herbicides.

However, such an approach caused the expose of soil surface to heavy rains and erosion. Furthermore, the intensive use of herbicides has led to changes in weed flora as species more tolerant to these herbicides become more dominant (Heap, 2012). For example, glyphosate is widely considered the world's most important herbicide because of its many desirable characters (Baylis, 2000). During the first years of use, no cases of resistant weeds were reported, and the evolution of glyphosate resistance was considered unlikely (Bradshaw et al., 1997). Unfortunately, glyphosate-resistant weeds have evolved and spread, currently about 20 weed species in several countries (Heap, 2012). Repeated use of glyphosate in no-tillage systems can greatly increase the risk of glyphosate resistance, even in the absence of glyphosate-resistant crops (Neve et al., 2003; Urbano et al., 2007; Owen et al., 2009). Less competitive weeds, such as annual grasses or legumes that are easily controlled, are progressively replaced by more aggressive and competitive weeds. These weeds compete hard with trees, reduce their yield and quality and may make harvesting difficult. Therefore, in our days a management rather than a total elimination of weed flora is desirable while maintenance of a balanced weed flora with less-competitive weeds could be of high potential importance and several examples will be discussed in the following sections. In addition, the wise use and clearly not the exclusion of herbicides such as glyphosate in orchards should be promoted in order to make its use sustainable and its efficacy high (Travlos and Chachalis, 2010).

3. Potentially Useful and Non-Competitive Natural Vegetation

Weeds with their presence protect the soil against water and wind erosion, which is especially important on slopes, where many orchards are planted. They also protect the soil against mechanical compaction, since bare soil is much more prone to compaction than soil covered by plants. *Polygonum aviculare* L. is an example of a plant which grows very well on compacted soil. Weeds, especially legumes, are sources of organic matter and nitrogen for the soil, which improves soil properties. Under that point of view, *Medicago arabica*, *M. polymorpha*, *Trifolium repens*, *T. pratense*, *Vicia villosa* and many more common and usually found in the orchards of Greece (Travlos, unpublished data) could have a crucial role and many beneficial effects.

Nutrients absorbed by weeds are returned to the soil and this is very important especially for perennial crops like trees.

Furthermore, weeds create a safe habitat for many beneficial organisms (even some harmful, too). Many of them also have medicinal value, such as *Equisetum arvense* L. and *Conyza canadensis* L. (Saluk-Juszczak et al., 2007; Cetojevic-Simin, 2010). Some weeds are indicators of the chemical or physical properties of the soil. For example, *Papaver rhoeas* L. and *Lathyrus tuberosus* L. indicate neutral or even basic soil pH. *Chenopodium album* L., *Amaranthus retroflexus* L. and many other weeds indicate soils rich in nitrogen. Many weeds are already used as human or domestic animal food and they can grow under extremely unfavourable conditions, under which cultivated plants cannot be grown (Lipecki, 2006). Weeds such as *Amaranthus* sp., *Sonchus oleraceus*, *Portulaca oleracea* belong to this category. In general, weeds preserve the biological balance in orchards. The presence of vegetative cover provides habitats to encourage biodiversity. The wildlife encouraged will include predators of insect pests, which would otherwise have to be controlled chemically. Some weeds can be used as phytoremediants to collect and remove toxic substances such as heavy metals from the environment (Baker and Brooks, 1989; Reeves and Baker, 2000).

Now that the most negative and positive aspects of weeds in orchards are well known, most farmers tend to let some weeds grow freely in their orchards as long as they do not significantly harm their trees. Weeds can be controlled in many ways, depending mainly on the financial resources of the grower, while environmental protection is more often taken into account. An alternative approach to the use of herbicides is the use of cover crops. Previous research recommended the use of live winter mulch in orchards that is killed in early spring, either naturally or by herbicides (Atkinson et al., 1980; Skroch and Shribbs, 1986). Cover crops residues on the soil surface reduce soil erosion, nutrient leaching and runoff, damages soil structure and organic matter content, and suppress weeds (Barnes and Putnam, 1983; Putnam and DeFrank, 1983; Mohler, 1991; Swanton and Murphy, 1996). Planting cover crops, such as clovers, grass or other plants can be an important step in turning fragile or marginal lands into productive orchards. By slowing the spread of weeds, cover crops help seedlings compete for sunlight, nutrients and water, with reduced need for herbicides or other weed control. Some cover crops also enrich and stabilize the soil. The root systems of cover crops help prevent erosion by securing the soil. They are particularly useful on slopes where rain can carry soil into waterways. Depending on the species; they can help loosen compacted soil and increase soil organic matter, moisture and fertility.

A well-managed orchard floor cover between the tree rows also provides a stable surface upon which machinery can be operated under wet conditions that otherwise would prevent access to the orchard. Cover crops should be mowed at least once a year. This reduces habitat for rodents that might feed on tree seedlings (Harrington et al., 1999; 2005).

The effects of several annual legume cover crops (annual medics, berseem clover and medium red clover) on weed populations have already been studied (Fisk et al., 2001). Density of winter annual weeds were between 41 and 78% lower following most cover crops when compared with no cover control, while dry weight was between 26 and 80% lower in all sites. There are a number of mechanisms responsible for the effect of cover crops on weeds. The living cover crop can reduce light (Teasdale and Mohler, 1993). Weeds attempting to establish along with a cover crop would be in competition for resources and may not develop sufficiently to survive the winter. Cover crop residue can modify the conditions under which weeds germinate or regrow in the spring. Such effects could be due to changes in soil temperature, increase in soil moisture, release of allelopathic chemicals, and physical impediments to weed seedlings (Facelli and Pickett, 1991; Teasdale and Mohler, 1993; Teasdale, 1996).

Regarding allelopathy, many legumes produce alkaloids that may be responsible for the suppressive effects on a wide range of weeds (Fujii, 1999). Hence, several legumes have the potential to be used for biologically based weed control methods in organic cropping systems. Indeed, the most practical and immediate way to use allelopathy in weed control is to use allelopathic cover crops in or apply residues of allelopathic weeds or crops as mulches (Rice, 1974; Caamal-Maldonado et al., 2001; Dhima et al., 2007).

It has to be noted that in some cases there are significant differences, not only among species but also among cultivars of the same species used as cover crops, regarding their ability to suppress weed growth. Field trials were conducted in Canada to measure the growth of brown mustard (*Brassica juncea*), in mowed and nonmowed production, as influenced by alsike (*Trifolium hybridum*), balansa (*T. michelianum*), berseem (*T. alexandrinum*), crimson (*T. incarnatum*), Persian (*T. resupinatum*), red (*T. pratense*), and white Dutch (*T. repens*) clover and rye (*Secale cereale*). Although the effects varied with location, timing, and species, the characteristics of berseem clover (upright growth, long stems, high biomass, and late flowering) would support its use as a cover crop or forage in similar regions (Ross et al., 2001). Annual medics and red clover reduce density and dry weight of winter and summer annual weeds.

Under dryland conditions, large-seeded legumes and sweetclover (*Melilotus officinalis*) grown as green fallow provided better weed suppression than *Trifolium* spp. (Schlegel and Havlin, 1997). Research suggests that competitive abilities vary among clover species and cultivars within a species and with intent of use. Nelson et al. (1991) observed less weed biomass with an annual clover (crimson) than with two perennial clovers (red and white) used as spring cover crops.

Previous research on legumes and other species, investigating the possibility of biological weed control through plant interference has focused on the use of winter annual or perennial species sown in the fall and suppressed or killed with a herbicide application or mechanical disturbance in the spring (Enache and Ilnicki, 1990; Grubinger and Minotti, 1990; Mohler, 1991; Eberlein et al., 1992; Curran et al., 1994). Species evaluated include alfalfa *(Medicago sativa)*, pea, hairy vetch and several clovers (crimson, ladino, white and subterranean). Hairy vetch, ladino clover, subterranean clover, and white clover show promise, but none of these species is consistently cold tolerant, and spring growth of these species can deplete soil water resources. Therefore, specialized cover or mulch species selected to suppress weeds without reducing main crop yield are widely adopted and known as smoother plants (De Haan et al., 1994). Smother plants include several legumes (such as annual medic) and have the potential to suppress weed growth (De Haan et al., 1994), increase soil water infiltration, decrease soil erosion (Cripps and Bates, 1993), contribute N to the main crop (Decker et al., 1994), and reduce economic risk.

In the bottom of the line, research investigating the influence of soil type, precipitation, weed species, and soil nutrient status on smother plant performance will be necessary to identify the environments in which several legumes as smother plants can be successfully integrated into cropping systems (De Haan et al., 1997).

4. Challenges and Trends of Integrated Orchard Floor Management

Integrated weed management involves the use of multiple strategies to manage weed populations in a manner that is economically and environmentally sound. Such strategies include cultural, mechanical, chemical, and biological methods (O'Donovan, 1997; Anderson, 2007).

As previously mentioned, orchards may be infested with a variety of annual and perennial weeds, each competing with the trees for water and nutrients. Competition for these resources is of greater concern with young trees because weeds can significantly reduce their growth and consequently their production. Moreover, special attention should be paid in the summer period. Integrated weed management strategies vary from orchard to orchard and are influenced by location in the state, climatic conditions, soil texture and profile, irrigation practices, topography, cost, and grower preferences. There are several components to a good orchard weed management program. These include preventive strategies, orchard floor management, and weed monitoring.

Weeds on the orchard margins should not be ignored since they produce seeds that disperse into the orchards and their presence there is the first step of their invasive nature (Blumenthal and Jordan, 2001). Therefore, aggressive or noxious weeds need to be controlled before they set seeds. In addition, in the case of perennial weeds propagated with vegetative organs (rhizomes, bulbs, tubers etc.) the tires of vehicles and equipment should be cleaned before entering a "clear" orchard because seeds and reproductive parts of weeds can be transported along with them. Weed monitoring, and especially detection of new weeds and weeds that escaped previous control treatments is a key component of a weed management plan. Moreover, weeds in tree rows must be managed, but annual weeds between rows may be beneficial as cover crops and therefore their growth and dispersal should be encouraged (Travlos, unpublished data).

The potential of allelopathy for weed control in agriculture has for long been controversial, because evidence for direct and strong allelopathic effects and ecological significance is often difficult to prove and use (Blum et al., 1999; Inderjit and Keating, 1999). Nevertheless, crop plants with superior weed suppressive ability under field conditions would be highly desirable in agriculture, since they may reduce both number and population size of other species (Chaves and Escudero, 1997; Olofsdotter, 1998). Many allelopathic cover crops, mostly leguminous plants, provide protein rich food, and grow easily without artificial fertilizers, herbicides, insecticides and fungicides. In this regards, these allelopathic cover crops could save food shortage in rural area, and are useful for environmental conservation.

Hairy vetch (*Vicia villosa*) is considered one of the most promising cover plants for the weed control in orchards and even for landscape amendment in abandoned fields, since it shows strong allelopathic effects on other weeds (Fujii, 2003).

Another noticeable case is that of *Oxalis* spp. Bermuda buttercup (*Oxalis pes-caprae* L.) is a perennial small herb, member of Oxalidaceae family. It came from South Africa to Europe in 19th century and now is one of the most noxious and invasive weeds of Medditeranean regions and other environments (Vilà et al., 2006; Castro et al., 2007). In Greece it is one of the most dominant weed in the olive groves and vineyards (Damanakis and Markaki, 1990), while it often forms dense mats in any area with open space. This high competitive ability has to be investigated and further exploited, as long as *O. pes-caprae* has beneficial impacts by causing soil enrichment and stabilization of semi-stable areas and orchards and suppressing other weeds (Paspatis, 1987; Travlos et al., 2008). It has to be noted that this plant is already used as inter-row cover crop in citrus and olives in the Mediterranean countries (Castro et al., 2007; Travlos et al., 2008).

Reduction in weed population when several plant species are used as cover crops in orchards has been reported (Smeda and Putnam, 1988; Calkins and Swanson, 1995; Sainju, 1997). The physical presence of cover crops mulch on the soil surface can greatly reduce weed density and biomass (Putnam and DeFrank, 1983; Masiunas et al., 1995). Likewise, allelochemicals arising directly or indirectly from mulch or residue suppress weeds (Barnes and Putnam, 1986; Brown and Morra, 1995). Increasing levels of biomass with uniform soil coverage enhances weed suppression (Teasdale et al., 1991; Teasdale, 1996; Vidal and Bauman, 1996). Soil in orchards can be protected by cover crops consisting of annuals such as *Poa annua* L., perennials including grasses such as *Lolium perenne* L. and *Agropyron repens* L. and also some dicotyledons (members of Fabaceae, Asteraceae etc). However, further research for proper cover crops management is warranted to enhance weed suppression by them. In general, resident vegetation is generally permitted to grow in the areas between the tree rows but it must be managed through repeated mowing, tillage, or chemical treatment. Moreover, the presence of weeds difficult to control and quickly invasive to tree rows or neighbouring orchards such as *Conyza* spp. should be avoided (Travlos and Chachalis, 2010).

For ground covers to be a viable alternative to sprayed bare soil strips, an efficient mowing system is required within the tree rows since competition needs to be kept away from trees during the first few years of establishment. However, once trees are established, ground covers appear not to affect tree growth (Harrington et al., 1999; 2005). For species as low-growing as the dichondra (*Dichondra micrantha*), selective herbicides are necessary to stop them being taken over by more aggressive species.

Consequently, the use of glyphosate to keep soil bare at least between tree rows is also a realistic and cheap treatment. Seed costs for cover crops like fescue (*Festuca rubra*) and dichondra are high (Harrington et al., 1999; 2005). Therefore, the species that are already part of natural vegetation should be strategically encouraged (Travlos, unpublished data). Properly managed species can prevent invasion of the orchard by weeds that cause serious problems. Tall cover crops or weeds, however, increase the risk of frost damage in spring and should be mowed to reduce any risks.

Conclusion

Maintenance of a balanced flora, which precludes the dominance of aggressive species and includes some less-competitive or even useful weed species, seems an ideal situation for perennial crops. In particular, a controlled presence of non-competitive or even useful weeds maintains the balance of the weed flora and prevents invasion of more competitive species since bare ground for them to colonize is less available, while allelopathic effects also exist. For ground covers to be a viable alternative to sprayed or tilled bare soil, an efficient mowing system is required within the tree rows in order to avoid intense competition in the dry periods of the years and to prevent the seed production of taller and undesirable weed species. Future research should focus on legumes or other herbaceous species present inside or around orchards or other perennials which can suppress weeds. Such an integrated management should be clearly promoted and further studied in order to avoid the extended use of herbicides or the often mechanical cultivation and their negative effects.

References

Anderson, R. L. (2007) Managing weeds with a dualistic approach of prevention and control. *Agronomy Sustainable Development*. 27: 13–18.

Atkinson, D., Abernathy, W. and Crisp, C. M. (1980) The effect of several herbicides on moss establishment in orchards. *Proceedings of 1980 Brighton Crop Protection Conf., Weeds* p. 297–302.

Baker, A. J. M. and Brooks, R. R. (1989) Terrestrial higher plants which hyperaccumulate metallic elements: a review of their distribution, ecology and phytochemistry. *Biorecovery* 1: 81–126.

Barnes, J. P. and Putnam, A. R. (1983) Rye residues contribute weed suppression in no-tillage cropping systems. *Journal of Chemical Eco*logy 9: 1045–1057.

Barnes, J. P. and Putnam, A. R. (1986) Evidence for allelopathy by residue and aqueous extracts of rye. *Weed Science* 34: 384–390.

Baylis, A. D. (2000) Why glyphosate is a global herbicide: strengths, weaknesses and prospects. *Pest Management Science* 56: 299–308.

Blum, U., Shafer, S. R. and Lehmen, M. E. (1999) Evidence for inhibitory interactions involving phenolic acids in field soils: concepts vs. experimental model. *Critical Review of Plant Science* 18: 673–693.

Blumenthal, D. and Jordan, N. (2001) Weeds in field margins: a spatially explicit simulation analysis of canada thistle population dynamics. *Weed Science* 49 (4): 509–519.

Bradshaw, L. D., Padgette, S. R., Kimball, S. L., and Wells, B. H. (1997) Perspectives on glyphosate resistance. *Weed Technology* 11: 189–198.

Brown, S. M. and Whitwell, T. (1988) Influence of tillage on horseweed (*Conyza canadensis*). *Weed Techno*logy 2: 269–270.

Brown, P. D. and Morra, M. J. (1995) Glucosinolate-containing plant tissues as bioherbicides. *Journal of Agricultural and Food Chemistry* 43: 3070–3074.

Bruce, J. A. and Kells, J. J. (1990) Horseweed (*Conyza canadensis*) control in notillage soybeans (*Glycine max*) with preplant and preemergence herbicides. *Weed Techno*logy 4: 642–647.

Buhler, D. D. (1992) Population dynamics and control of annual weeds in corn (*Zea mays*) as influenced by tillage systems. *Weed Science* 40: 241–248.

Caamal-Maldonado, J. A., Jiménez-Osornio, J. J., Torres-Barragán, A., and Anaya, A. L. (2001) The use of allelopathic legume cover and mulch species for weed control in cropping systems. *Agronomy Jounal* 93: 27–36.

Calkins, J. B. and Swanson, B. T. (1995) Comparison of conventional and alternative nursery weed management strategies. *Weed Techno*logy 9: 761–767.

Castro, S., Loureiro, J., Santos, C., Ater, M., Ayensa, G., and Navarro, L. (2007) Distribution of flower morphs, ploidy level and sexual reproduction of the invasive weed *Oxalis pes-caprae* in the Western area of the Mediterranean region. *Annals of Botany* 99: 507–517.

Cetojevic-Simin, D. D. (2010) Antioxidative and antiproliferative activities of different horsetail (*Equisetum arvense* L.) extracts. *Journal of Medicinal Food* 13 (2): 452–459.

Chaves, N. and Escudero, J. C. (1997_ Allelopathic effect of *Cistus ladanifer* on seed germination. *Functional Ecology* 11: 432–440.

Christensen, S. (1995) Weed suppression ability of spring barley varieties. *Weed Research* 35: 241–247.

Crips, R. W. and Bates, H. K. (1993) Effect of cover crops on soil erosion in nursery aisles. *Weed Technol*ogy 5: 664–673.

Curran, W. S., Hoffman, L. D. and Werner, E. L. (1994) The influence of a hairy vetch (*Vicia villosa*) cover crop on weed control and corn (*Zea mays*) growth and yield. *Weed Technol*ogy 8: 777–784.

Damanakis, M. and Markaki, M. (1990) Studies on the biology of *Oxalis pes-caprae* L., under field conditions in Crete. *Zizaniology* 2: 145–154.

Decker, A. M., Clark, A. J., Meisinger, J. J., Mulford, F. R., and McIntosh, M. S. (1994) Legume cover crop contributions to no-tillage corn production. *Agronomy Journal* 86: 126–135.

De Haan, R. L., Wyse, D. L., Ehlke, N. J., Maxwell, B. D., and Putnam, D. H. (1994) Simulation of spring-seeded smother plants for weed control in corn (*Zea mays*). *Weed Science* 42: 35–43.

Dhima, K. V., Lithourgidis, A. S., Vasilakoglou, I. B., and Dordas, C. A. (2007) Competition indices of common vetch and cereal intercrops in two seeding ratio. *Field Crops Research* 100: 249–256.

Didon, U. M. E. and Hansson, M. L. (2002) Competition between six spring barley (*Hordeum vulgare* ssp. *vulgare* L.) cultivars and two weed flora in relation to interception of photosynthetic active radiation. *Biological Agriculture and Horticulture* 20: 257–274.

Eberlein, C. V., Sheaffer, C. C. and Oliveira, V. F. (1992) Corn growth and yield in an alfalfa living mulch system. *Journal of Production Agriculture* 5: 332–339.

Enache, A. J. and Ilnicki, R. D. (1990) Weed control by subterranean clover (*Trifolium subterraneum*) used as a living mulch. *Weed Technology* 4: 534–538.

Facelli, J. M. and Pickett, S. T. A. (1991) Plant litter: Its dynamics and role of legumes in effects on plant community structure. *Botanical Review* 57: 1–32.

Fisk, J. W., Hesterman, O. B., Shrestha, A., Kells, J. J., Harwood, R. R., Squire, J. M., and Sheaffer, C. C. (2001) Weed suppression by annual legume cover crops in no-tillage corn. *Agronomy Journal* 93: 319–325.

Fujii, Y. (1999) Allelopathy of hairy vetch and *Mucuna*; their application for sustainable agriculture. p. 289–300. In: C. H. Chou, et al. (ed.) *Biodiversity and allelopathy from organisms to ecosystems in the pacific.* Academia Sinica, Taipei.

Fujii, Y. (2003) Allelopathy in the natural and agricultural ecosystems and isolation of potent allelochemicals from Velvet bean (*Mucuna pruriens*) and Hairy vetch (*Vicia villosa*). *Biological Scientific Space* 17 (1): 6–13.

Grubinger,V. P. and Minotti, P. L. (1990) Managing white clover living mulch for sweet corn with partial rototilling. *American Journal of Alternative Agriculture* 5: 4–12.

Harrington, K., Zhang, T., Osborne, M., and Rahman, A. (1999) Orchard weed control from *Dichondra micrantha* ground covers. *Australian Weeds Conference.* 12: 250–254.

Harrington, K. C., Hartley, M. J., Rahman, A., and James, T. K. (2005) Long term ground cover options for apple orchards. *New Zealand Plant Protection* 58: 164–168.

Heap, I. (2012) International Survey of Herbicide Resistant Weeds. *http://www.weedscience.com.* Accessed: July 2012.

Horowitz, M. (1973) Competitive effects of three perennial weeds, *Cynodon dactylon* (L.) Pers., *Cyperus rotundus* L. *and Sorghum halepense* (L.) Pers., on young citrus. *Journal of Horticulturl Science* 48 (2): 135–147.

Inderjit and Keating, K. I. (1999) Allelopathy: Principles, Procedures, Procecesses, and Promises for Biological Control. pp. 141-231. In: *Advances in Agronomy*. Sparks, D. L. (ed.), Academic Press.

Jordan, N. (1993) Prospects for weed control through crop interference. *Ecological Applicatioc* 3: 84–91.

Lipecki, J. (2006) Weeds in orchards – pros and contras. *Journal of Fruit and Ornamental Plant Research* 14: 13–18.

Masiunas, J. B., Weston, L. A. and Weller, S. C. (1995) The impact of rye cover crops on weed populations in a tomato cropping system. *Weed Science* 43: 318–323.

Mohler, C. L. (1991) Effect of tillage and mulch on weed biomass and sweet corn yield. *Weed Technology* 5: 545–552.

Nelson, W. A., Kahn, B. A. and Roberts, B. W. (1991) Screening cover crops for use in conservation tillage systems for vegetables following spring plowing. *HortScience* 26: 860–862.

Neve, P., Diggle, A. J., Smith, F. P., and Powles, S. B. (2003) Simulating evolution of glyphosate resistance in *Lolium rigidum*, II: past, present and future glyphosate use in Australia cropping. *Weed Research* 43: 418–427.

O'Donovan, J. T. (1997) Integrated weed management in reduced tillage systems. *Proceedings from the 9th Annual Meeting of the Saskatchewan Soil Conservation Association*, Saskatoon, Canada, 9 pp.

Olofsdotter, M. (1998) Allelopathy for weed control in organic farming. *Sustainable Agriculture for Food, Energy and Industry* 453–457.

Owen, M. J., Owen, R. K. and Powles, S. B. (2009) A survey in the Southern belt of Western Australia did not find *Conyza* spp. resistant to glyphosate. *Weed Technology* 23: 492–494.

Owen, M. J. and Powles, S. B. (2009) Distribution and frequency of herbicide-resistant wild oat (*Avena* spp.) across the Western Australian grain belt. *Crop Pasture Science* 60: 25–31.

Paspatis, E. A. (1987) Chemical, cultural and biological control of *Oxalis pes-caprae* in vineyards in Greece. Proceedings of a Meeting of the EC Experts' Group, Commission of the European Communities, *Weed Control on Vine and Soft Fruits*. Dublin Ireland 12–14 June, 1987, pp: 27–29.

Putnam, A. R. and DeFrank, J. (1983) Use of phytotoxic plant residue for selective weed control. *Crop Protection* 2: 173–181.

Reeves, R. D. and Baker, A. J. M. (2000) Metal accumulating plants, pp. 193–229. In: Raskin, I. and B. Ensley (eds.) *Phytoremediation of Toxic Metals: Using Plants to Clean up the Environment*. New York: Wiley.

Rice, E. L. (1974) *Allelopathy*. Academic Press, New York.

Ross, S. M., King, J. R., Izaurralde, R. C., and O'Donovan, J. T. (2001) Weed suppression by seven clover species. *Agronomy Journal* 93: 820–827.

Sainju, U. M. (1997) Winter cover crops for sustainable agriculture systems. *Horticultural Science* 2: 21–28.

Saluk-Juszczak, J., Olas, B., Pawlaczyk, I., Gancarz, R., and Wachowicz, B. (2007) Effects of the extract from *Conyza canadensis* on human blood platelet aggregation. *General Physiology* and *Biophysics* 26 (2): 150–152.

Schlegel, A. J. and Havlin, J. L. (1997) Green fallow for the central Great Plains. *Agronomy Journal* 89: 762–767.

Shields, E. J., Dauer, J. T., VanGessel, M. J., and Neumann, G. (2006) Horseweed (*Conyza canadensis*) seed collected in the planetary boundary layer. *Weed Science* 54: 1063–1067.

Shui, J. G., Liao, G. G., Au, J., Zhou, Q. K., and Allard, J. L. (2004) Effects of different natural vegetation management measures on red soil erosion in hilly orchards. *ISCO 2004 – 13th International Soil Conservation Organisation Conference – Brisbane, Country?,* July 2004 Conserving Soil and Water for Society: Sharing Solutions.

Skroch, W. A. and Shribbs, J. M. (1986) Orchard floor management: An Overview. *Horticultural Science* 21: 390–394.

Smeda, R. J. and Putnam, A. R. (1988) Cover crop suppression of weeds and influence on strawberry yields. *Horticultural Science* 23: 132–134.

Swanton, C. J. and Murphy, S. D. (1996) Weed Science beyond the weeds: The role of integrated weed management (IWM) in agroecosystem health. *Weed Science* 44: 437–445.

Teasdale, J. R. (1996) Contribution of cover crops to weed management in sustainable agricultural systems. *Journal of Production Agriculture* 9: 475–479.

Teasdale, J. R., Beste, C. E. and Potts, W. E. (1991) Response of weeds to tillage and cover crop residue. *Weed Science* 39: 195–199.

Teasdale, J. R. and Mohler, C. L. (1993) Light transmittance, soil temperature, and soil moisture under residue of hairy vetch and rye. *Agronomy Journal* 85: 673–680.

Travlos, I. S. and Paspatis, E. (2008) Aspects of the allelopathic effects of heliotrope (*Heliotropium europaeum*) on *Avena sativa*, *Phaseolus vulgaris* and *Spirodela polyrhiza*. *Allelopathy Journal* 21: 397–404.

Travlos, I. S., Paspatis, E. and Psomadelli, E. (2008) Allelopathic potential of *Oxalis pes-caprae* tissues and root exudates as a tool for integrated weed management. *Journal of Agronomy* 7 (2): 202–205.

Travlos, I. S. and Chachalis, D. (2010) Glyphosate-resistant hairy fleabane (*Conyza bonariensis*) is reported in Greece. *Weed Technology* 24: 569–573.

Travlos, I. S., Economou, G. and Kanatas, P. J. (2011) Corn and barnyardgrass competition as influenced by relative time of weed emergence and corn hybrid. *Agronomy Journal* 103: 1–6.

Urbano, J. M., Borrego, A., Torres, V., Leon, J. M., Jimenez, C., Dinelli, G., and Barnes, J. (2007) Glyphosate-resistant hairy fleabane (*Conyza bonariensis*) in Spain. *Weed Technology* 21: 396-401.

Vidal, R. A. and Bauman, T. T. (1996) Surface wheat residues, giant foxtail and soybean yield. *Weed Science* 44: 939–943.

Vilà, M., Bartomeus, I., Gimeno, I., Traveset, A., and Moragues, E. (2006) Demography of the invasive geophyte *Oxalis pes-caprae* across a Mediterranean Island. *Annals of Botanty* 97: 1055–1062.

Weaver, S. E. (2001) The biology of Canadian weeds. 115. *Conyza canadensis*. *Canadian Journal of Plant Science* 81: 867–875.

Young, B. G. (2006) Changes in herbicide use patterns and production practices resulting from glyphosate-resistant crops. *Weed Technology* 20: 301–307.

In: Weeds and Their Ecological Functions
Editor: Alireza Taab
ISBN: 978-1-62808-332-3

Chapter 5

THE ROLE OF COVER CROPS IN AGRO-ECOSYSTEMS MANAGEMENT

Rosario P. Mauro, Gaetano Roberto Pesce and Giovanni Mauromicale*
Dipartimento di Scienze delle Produzioni Agrarie e Alimentari (DISPA), University of Catania, via Valdisavoia, Catania, Italy

ABSTRACT

Cover cropping has been a known technique since ancient times and has been used in civilizations that are distant from each other, in both space and time, as a tool to improve soil fertility and to ensure yield stability. Over recent years, the growing side effects of intensive agriculture have stimulated a renewed attention toward cover crops, because of their ability to reduce the environmental impact deriving from the increased use of agrochemicals and agricultural mechanization. This chapter provides an overview of cover cropping typologies and their significance in different agricultural contexts, including their effects on physical (structure, porosity, thermal regime, water evaporation, erodibility) and chemical (organic matter content, nitrogen, phosphorus, potassium and microelements) soil characteristics, as well on the

* Corresponding author: Giovanni Mauromicale, Professor, Dipartimento di Scienze delle Produzioni Agrarie e Alimentari (DISPA), University of Catania, via Valdisavoia, 5 – 95123 Catania (Italy). Tel: +39095234409; Fax: +39095234449; E-mail: g.mauromicale @unict.it.

biocenotic components of the agroecosystems (weeds, pathogens, arthropods, nematodes). The main characteristics of the better-suited species for cover cropping purposes are also reported. Among these characteristics, particular attention has been paid to soil and climate preferences, vegetative vigour, persistence over time or competitive ability against weeds, all traits considered of pivotal importance in determining their possible diffusion as cover crops in agricultural systems. Finally, it is worth considering that cover crops performance improvement is critically affected by agricultural (i.e. adopting measures aimed at cost reduction, identification of seed mixtures suitable for enhancing the biological complementarity among species) and genetic innovation (breeding of specific genotypes for cover cropping purposes).

1. INTRODUCTION

The intensification of cultivation practices, particularly relevant from the second post-war period, has brought with it numerous side-effects, including the depletion of natural resources, the reduction of soil fertility and the accumulation of agricultural pollutants in food and environment. Over the last decades, this has highlighted the need to seek new agronomical solutions that may give a more sustainable profile to farming systems. In this context, there has been a renewed interest in cover cropping, since it can provide multiple benefits to farming systems from an agronomical and ecological point of view (Lu et al., 2000). Commonly, the term cover cropping refers to the mono- or intercropping of herbaceous plants (cover crops) over either part of a year, or over an entire year, in order to improve the performance of the cash crops (Campiglia, 1999; Caporali and Campiglia, 2001). Although the term cover crop was used for the first time in its current meaning only at the end of the 19th century, cover cropping has deep roots in human history and was present in civilizations distant from each other in space and in time. The fourth book of *Li Ki* or *Book of Rites*, written under Han Dynasty (206 B.C.-220 A.D.), reports that use of weeds and grasses as green manure was known in China under Zhou Dynasty (from XII to III century B.C.) (Pieters, 1927). Theophrastus (370-285 B.C.), in his *Enquiry into Plants*, testifies that Ancient Greeks knew of the agronomic value of fava bean to promote soil fertility. Columella, in his *De re rustica* (about 65 B.C.), wrote about the usefulness of lupine which, despite its poor commercial value, has a low demand for cultivation practices and enhances both the soil fertility and nutritional status of weak grapevines.

Nowadays, cover cropping is a world-wide practice both in herbaceous and orchard cropping systems, through which positive effects can be reached in improving soil fertility, reducing soil erosion and managing weeds and pests in farming systems (Ingels and Klonsky, 1998).

However, these positive effects arise from an appropriate choice of cover crops to use, which, in turn, is the result of an adequate knowledge of their biological behavior and the most proper techniques to manage them.

2. Cover Crop Typologies and Their Management

Cover crops can be employed both in orchard and herbaceous cropping systems. In the former, they are intercropped with the cash crop and used as living mulch, while in the latter case their presence is often temporally separated from the cash crop, giving their main benefits either as green manure or as dead mulch (Caporali and Campiglia, 2001). Cover crop typologies vary greatly according to the utilized species, field duration, occupied field and used management techniques. When the natural flora is utilized (natural cover cropping), cover cropping can be permanent or temporary. In the first case weeds are left to grow freely and then managed with appropriate agronomic interventions.

In a temporary cover cropping system, the cover crop is maintained only in those periods when there is less competition for water between the cash crop and the cover crop (McGourthy and Christensen, 1998). The so-called artificial cover cropping is done by seeding one or more selected species, and it may be either permanent or temporary too. In this case, the species must be chosen taking into account the expected benefits, as well as species' characteristics (water use, competition, ease of establishment and maintenance, rapidity in soil covering). When the cover crop is intercropped with the cash crop (as in orchards), cover cropping can be either total or partial. In the first case, the cover crop covers the entire cultivated area (both rows and inter-rows).

This generally tends to increase the water and nutrient competition among the cash crop and the cover crop, especially in those areas characterized by uneven rainfall distribution and high evaporative demands. For this reason, this type of floor management could find useful application in order to limit the excessive vegetative vigor of the cash crop, so balancing the source/sink ratio and creating more suitable microclimatic conditions for fruit ripening (Lopes et al., 2011).

In the second case, the cover crop is limited to inter-row strips, in order to reduce water and nutrient competition toward the cash crop. This system may have useful applications in drought areas, whose soils are periodically exposed to run-off.

In most cases, cover crop management systems should aim to minimize water and nutrient competition between the cash crop and cover crop. This can be achieved by means of periodical cuttings of the aboveground biomass (when the used species are able to produce tillers or offshoots), followed by biomass removal or chopping (Campiglia and Caporali, 1996). In the case of self-reseeding cover crop species, such intervention might be made some time before the flowering period, in order not to endanger the sward formation in the following season.

3. Cover Crops Effects on Agro-Ecosystems

Cover crops affect many aspects of agro-ecosystems, by modifying:

- The physical soil characteristics and its susceptibility to erosion;
- The chemical soil characteristics and mineral nutrition of plants;
- The biocenotic component present within the agroecosystems.

3.1. Effects of Cover Crops on Physical Soil Characteristics and Erosion

3.1.1. Soil Structure and Porosity

Cover crops are known to promote and protect soil structure, mainly thanks to their own biomass as well as to the release into the soil of polysaccharide substances. The release of biomass involves an increase of soil organic matter, which helps to stabilize soil aggregates. Moreover, the aboveground biomass provides protection toward the soil structural aggregates, through the dissipation of the kinetic energy of rain drops. The protection of the aggregates is also useful in order to reduce the risk of soil crusting.

Moreover, the growth and subsequent decomposition of the cover crop root system, helps to increase the soil porosity, so promoting a better air/ water ratio within the arable layer and a faster root growth of the subsequent herbaceous cash crop (Breland, 1995).

3.1.2. Soil Thermal Regime and Water Evaporation

The dead mulch of crop cover left on soil surface, intercepts the incident solar radiation and constitutes a barrier to gas exchange between the soil and atmosphere. This involves an attenuation of the upper soil layer temperature, as well as a reduction of water evaporation from its surface (Bond and Willis, 1969; Ashworth and Harrison, 1983; Wagner-Riddle et al., 1994). Research conducted by Teasdale and Abdul-Baki (1995), concluded that the burial of residues of *Vicia villosa* Roth. within the top 0-5 cm soil layer, led to a 5 and 7 °C reduction, as compared to an unmulched and a mulched soil with opaque plastic nets, respectively. This provides evidence on the role that cover crops with mulching effects can play during the warm periods (e.g. in Mediterranean climate), when soil temperatures reach higher than optimal values for the growth of cultivated plants and the life of soil microorganisms, by reducing soil water evaporation and avoiding the rapid lowering of soil water potential.

3.1.3. Soil Erosion

Water erosion of soil is a natural geological phenomenon, particularly pronounced in soils with high silt and/or clay content, low softness, low content of organic matter, oxides and sesquioxides of Fe and Al and marked slopes (Cotching et al., 1979; Marks and Soane, 1987; Pagliai et al., 1995; Shanks et al.,1998; Batey, 2009). Vegetation is increasingly used to prevent soil erosion by runoff over the land surface, instead of engineering structures like bunds and terraces on sloping agricultural land. Cover crops are suggested as one of most effective tools to control soil erosion, as they ensure the dissipation of the kinetic energy of rain drops, the reduction of soil crusting, the increase in the soil softness and its water permeability to water, the reduction of run-off and the retention of soil particles (Rod, 1977; Langdale et al., 1991; Shanks et al., 1998; Morgan and Duzant, 2008). The best results are secured by fast-growing species, associated with dense vegetation and a well developed and branched root system. In Mediterranean environments, where rainfall occurs almost exclusively in the autumn-winter period, the use of cover crop with an autumn-winter growing period can help to significantly reduce water erosion. Among the Poaceae family, particularly effective are *Lolium perenne* L., *Festuca arundinacea* Schreb., *F. ovina* L. and *Bromus hordeaceus* L., also because they are self-reseeding (Adler, 1995; Shanks et al., 1998). *Secale cereale* L., *Hordeum vulgare* L. and *Avena sativa* L. also provide a good erosion control, but their duration is limited to one season only, as they are not able to re-seed (Caporali et al., 1990; McGourthy, 1994; De Baets et al., 2011).

Appreciable effects are also ensured by some Fabaceae species: *Trifolium subterraneum* L., *T. brachycalycinum* Katzn. et Morley and *T. hirtum* All., all provided with effective self-reseeding mechanisms, and some annual medics (*Medicago* spp.), characterized by an aboveground dry matter yield even higher than 4 t ha^{-1} (Shanks et al., 1998; Campiglia, 1999; Caporali and Campiglia, 2001). The Fabaceae species, however, seem to be less effective than the Poaceae, mainly because of the lesser extent of their root system. For this reason their use, either in pure culture or in plants mixture, is more appropriate when, together with the reduction of erosion, growers pursue an increase in the nitrogen content of the soil (Shanks et al., 1998).

3.2. Effects of Cover Crop on Chemical Soil Characteristics

3.2.1. Organic Matter Content

Organic matter released by cover crops provides the soil with several benefits, whose extent depends closely on the cover crop species and biomass management system (Baumeister, 1964; Porter, 1998; Warner, 1999). Regarding the former, the species with succulent tissues, high protein and simple carbohydrates content, as well as a low C/N ratio, mainly undergo the mineralization processes, so representing a relatively fast source of available nutrients for the successive or consociated cash crops when used as green manure (Lehman et al., 2000).

To this end, legumes are the most suitable species, thanks to their N-fixing ability at the roots level. *Lupinus angustifolius* L., *Medicago scutellata* (L.) Mill., *M. truncatula* Gaertn., *Melilotus officinalis* (L.) Pallas, *Trifolium alexandrinum* L., *T. incarnatum* L., *T. subterraneum* L., *Vicia villosa* Roth. and *V. faba* var. *minor* (Harz) Beck, are all capable of providing production of dry matter exceeding 3 t ha^{-1} (Cherr et al., 2006). On the contrary, the biomass of species with high C/N ratio and with high content of structural carbohydrates and lignin, is more persistent in soil and undergo mainly the humification process, so helping to improve the soil physical characteristics (structure, porosity, field capacity, etc.), and increasing the soil cation exchange capacity (Sarrantonio, 2000; Sullivan, 2003).

These species largely belong to the *Poaceae* family, such as *Secale cereale* L., *Avena sativa* L., *Lolium multiflorum* Lam., *Hordeum spp.* and *Triticum spp.* Moreover, they are more effective than legumes in capturing soil residual N, subtracting it to leaching phenomena during the rainy season, hence their suitability to be used as catch crops.

However, this ability must be carefully assessed in relation to the amount of N in the soil available for the main crop, as well as in relation to the growing period of this last, in order to avoid either the exacerbation of competition toward the main crop, or the unavailability of the element for the subsequent crops (Cherr et al., 2006). Cover cropping may also achieve the dual goal of improving the nutritional and physical characteristics of the soil by employing mixtures of species with complementary characteristics. This is the case, for example, of cover crop mixtures such as *Trifolium incarnatum* L. - *Secale cereale* L., *Pisum sativum* L. - *Secale cereale* L., *Vicia villosa* Roth. - *Triticum aestivum* L., *Vicia sativa* L. - *Avena sativa* L. (Griffin et al., 2000; Karpenstein-Machan and Stuelpnagel, 2000; Cline and Silvernail, 2001).

As regards the sward management, the biomass of the cover crop can be incorporated into the soil at full anthesis (green manure), or after seeds maturation, in order to allow the species to self-reseed (as in the case of many clover species) (Hirshfelt, 1998). However, it must be taken into account that a delay in mowing cover crops causes a greater lignification of plant tissues, an increase in C/N ratio, and a longer persistence of cover crop residues into the soil, thereby shifting the main functions performed by the organic matter into the soil (Hirshfelt, 1998; Sarrantonio, 2000).

3.2.2. Nitrogen Content

In nature, nitrogen is characterized by the scarcity of its forms that may be directly assimilated by plants, as well as the high mobility of its mineral forms within the soil. Nitrogen fertilization, therefore, is needed to maintain adequate yield levels, but at the same time represents a potential source of environmental impact due to denitrification (air pollution) and leaching (hydrosphere pollution) (Erisman et al., 2007; Ierna et al., 2012).

Through the careful choice of the cover crop species, it is possible to appreciably reduce mineral fertilizations, at the same time decreasing the nitrogen escape from the soil-plant-atmosphere system (Ingels et al., 1998; Fatima et al., 2012). Regarding the first aspect, numerous studies show that the green manure of legumes may satisfy all or most nitrogen demands of a given succeeding crop (Hargrove, 1986; Holderbaum et al., 1990; Power, 1990; Caporali and Campiglia, 1994; Bonari et al., 1997; Burket et al., 1997; Mitchell and Tell, 1997; Kuo and Jellum, 2002; Bergkvist et al., 2011; Salmeron et al., 2011), as nitrogen contribution from legumes cover crop can exceed 200 kg ha^{-1} $year^{-1}$ (Kuo and Sanjiu, 1998), varying according to the utilized species, climate conditions of the growing environment and sward management (Jansson and Persson, 1982; Frankerberger and Abdelmagid,

1985; Muller et al., 1988; Vigil and Kissel, 1991; Quemada and Cabrere, 1995; Campiglia et al., 2011, 2012). In mild winter climates, significant nitrogen inputs may be ensured by the green manure of *Medicago scutellata* (L.) Mill., *Trifolium incarnatum* L., *T. alexandrinum* L., *Vicia dasycarpa* Ten. (Lu et al., 2000). The amount of nitrogen biologically fixed by these species varies, between 50 and 200 kg ha^{-1}, of which a variable fraction between 40 and 60% may be directly available for the succeeding crop (Sullivan, 2003), being sufficient to satisfy a significant proportion of the nitrogen requirements of crops like corn or processing tomatoes (Decker et al., 1994; Abdul-Baki et al., 1997). It should be noted, however, that if the maximum nitrogen requirement of the succeeding cash crop falls in a different period to that of the most intense mulch mineralization, nitrogen could face relevant dispersion phenomena (Huntington et al., 1985; Wagger, 1989). In this case, the use of mixtures of legumes and grasses, if properly balanced, allows obtaining a higher C / N ratio and a more gradual nitrogen release from the residual organic matrix (Clark et al., 1994; Ranells and Wagger, 1997; Miguez and Bollero, 2005). Moreover, several grasses, used as catch crops, can effectively limit soil nitrogen losses; this is the case of *Secale cereale* L., *Avena sativa* L., *Lolium multiflorum* Lam. or *Hordeum vulgare* L., which can more than halve nitrogen losses from the soil (Evans, 1977; Glenn and Welker, 1989; Meisinger et al., 1991; Wiedenfeld et al., 1999). To this end, some Brassicaceae (*Brassica hirta* Moench., *B. juncea* (L.) Czern, *B. napus* L.) and Hydrophyllaceae (*Phacelia tanacetifolia* Benth, etc.) are also useful, which, in addition to mitigating the effects of leaching, provide easily mineralizable biomass, so the nitrogen contained in it can become readily available for the succeeding cash crop (Shennan, 1992; Wyland et al., 1995; Campiglia, 1999).

3.2.3. Availability of Phosphorus, Potassium and Microelements

Cover crops are able to significantly alter the dynamics of other nutrients in the soil. Moser (1942) first noted that incorporation of some legumes residues into the soil may double the amount of available phosphorus and assimilable potassium.

This effect is due to the ability of these crops to assimilate not easily accessible elements and to deliver them, once used as green manure, into more accessible forms. *Lupinus albus* L. is able to solubilize large amounts of phosphorus and microelements present in alkaline soils (Gardner et al., 1981, 1983). This is explained by the ability of this species to release, through root exudation, large quantities of citrate ion, which is capable of solubilizing mainly phosphorus, iron, aluminium and manganese (Gardner et al., 1983).

This ability increases when *Lupinus albus* L. is intercropped with *Brassica napus* L. (Little et al., 2004). Under conditions of zinc and iron deficiency, some grasses release into the soil specific compounds (phytosiderophores), which play an important role in solubilization of these elements, increasing their availability for the intercropped cash crops (Hirshfelt, 1998). Cover crops can also augment the availability of macro- and microelements through their absorption from deeper soil layers and their subsequent release in more superficial layers of the soil, after biomass mineralization (Lu et al., 2000). Thanks to this mechanism, Abdul-Baki et al. (1997) showed that cover cropping *Vigna unguiculata* (L.) Walpers, *Vicia dasycarpa* Ten. or *Hordeum vulgare* L. lead to significant increases in potassium (177 kg ha^{-1}), phosphorus (17 kg ha^{-1}), iron (11 kg ha^{-1}) and aluminium (9 kg ha^{-1}) in the uppermost soil layer.

3.3. Effects of Cover Crops on Biological Communities of Agro-Ecosystems

3.3.1. Weeds

Weeds represent one of the chief limiting factors affecting agricultural systems, since they trigger a number of mechanisms (competition for light, water and minerals, release of allelochemicals) which strongly affect crop physiology, growth and yield potential (Hoffman et al., 1996). Therefore, management methods need to be implemented that can balance high efficacy, low costs and low environmental impact (Hutchinson and McGiffen, 2000). If properly managed, cover crops can limit the presence of weeds in cropping systems, through antagonistic effects resulting (i) in a reduction of the weeds' seed bank, (ii) in a reduction in the number of weed seedlings able to emerge, (iii) in a delay in weed seeds germination and plantlets' emergence and, consequently, in an attenuation of their ability to compete with the cash crop (Den Hollander et al., 2007a, b).

Generally, these effects are more marked when fast-growing, high biomass-yielding cover crops are used (Putnam, 1990; Den Hollander et al., 2007a). The Poaceae family comprises many plant species characterized by high growth rates and rapid initial ground covering, thereby being able to effectively inhibit the germination of several weeds with small seeds, generally with a photoblastic response (Putnam and De Frank, 1982).

The inhibition is due to the interception of solar radiation, especially in the red region (~660 nm), so shifting the red: far red ratio, which is responsible for triggering the germination of many small-seeded weeds (Smith and Morgan, 1983). It should be noted, however, that the transmission of light through the vegetation, particularly when the cover crop density is not uniform, can be very heterogeneous, despite the apparent homogeneity of the cover crop (Teasdale and Mohler, 1993). Studies conducted by Teasdale (1996) have shown that the presence of about 3 t ha^{-1} residues of *Vicia villosa* Roth. or *Secale cereale* L. allows the transmission through the sward of more than 50% of the incident PAR. This explains the unsatisfactory weed control when seeding is not carefully executed or, in any case, when the cover crop mulch is not uniform. The reduction of soil temperature determined by the mulching effect of a cover crop, though it is not able to prevent weed germination per se, may cause a delay in weeds emergence, so conferring them a certain competitive disadvantage (Teasdale and Mohler, 1993).

In addition to physical effects, the mulch of some cover crops determines the inhibition of germination and growth of many weed species through the release of allelochemicals into the soil (Kruidhof et al., 2009). The intensity of these effects, attributed to a heterogeneous class of compounds (polysaccharides, aminoacids, fatty acids, sterols, enzymes, flavonoids, flavonones), are affected by many factors that are still not fully clarified but which certainly play a role the genotype and the organ of the target species, its physiological age, and the presence of any concomitant stress induced by the environment (Lovett and Jessop, 1982; Olesek and Jurzysta, 1987; Inderjit and Duke, 2003). There are many cover crop species that may determine significant allelopathic effects. They include some grasses [*Secale cereale* L., *Avena sativa* L., *Sorghum bicolor* (L.) ssp. *drummondii* (Nees ex Steud.) de Wet and Harlan], Fabaceae [*Melilotus officinalis* (L.) Pallas, *Pisum sativum* L., *Lens culinaris* Medik., *Trifolium* spp., *Vicia villosa* Roth.] and Brassicaceae [*Brassica nigra* (L.) WDJ Koch, *B. juncea* (L.) Czern., *B. rapa* L., *Raphanus sativus* L.] (Weston et al., 1989; White et al., 1989; Bradow and Connick, 1990; Enache and Ilnicki, 1990). In a research conducted by Putnam and De Frank (1982), the dead mulch of *Secale cereale* L. has reduced the aboveground biomass production in *Ambrosia artemisiifolia* L. by 43%, in *Setaria viridis* (L.) Beauv. by 80% and in *Portulaca oleracea* L. by 100%. In a research conducted by Schilling et al. (1985), the dead mulch of *S. cereale* L. caused a severe reduction of the aboveground biomass of *Chenopodium album* L., *Amaranthus retroflexus* L. and *Ambrosia artemisiifolia* L. Among the Brassicaceae, *Brassica nigra* (L.) W.D.J. Koch and *B. juncea* (L.) Czern. are particularly effective in reducing germination and growth of weeds species, due to the high glucosinolates content in their green tissues (Bialy et al., 1990).

The allelopathic effects of some Fabaceae are due to the presence in their radical excreta of the aminoacid alanine, which is capable of interfering with the growth of several grasses (Schenk and Werner, 1991).

3.3.2. Plant Pathogens

3.3.2.a. Cover Crop-Plant Pathogens Interactions

Cover crops are known to enrich and diversify the biocenosis inhabiting agroecosystems, often resulting in significant limiting-effects on the activity of plant pathogens (Sullivan, 2003). It should be noted, however, that these effects do not always imply an improved health status of the cash crop, because of the complex interactions established between cover crop, main crop and plant pathogens (Costello and Daane, 1998).

In general, cover crops help the main crop to maintain those barriers (cuticles, waxes, antagonistic microorganisms present on leaf surface) that hamper the entry of certain pathogens, protecting the main crop from wind (against both mechanical stresses and transport of soil abrasive particles) or from the debilitating effect of some cultural practices (decreased transit of mechanical machineries for farming operations). The presence of a living or dead mulch on soil surface, in addition, may help to reduce the dissemination of fungal spores by rain drops (Phatak, 1998).

Moreover, many cover crops host, on their vegetation, populations of antagonistic microorganisms (bacteria, actinomycetes) that, once migrated on the vegetation of the cash crop, play a significant antagonistic effect against several plant pathogens (Phatak, 1998).

At soil level, the increase in organic carbon content, contributes to boost the biological complexity of soil microbiota communities, so decreasing the attack severity of many soil-borne pathogens (Lumsden et al., 1983). Emblematic here is the reduction of attacks by *Phytophthora parasitica* Dastur and *Pyrenochaeta lycopersici* Schneider and Gerlach found in tomato crops in response to the increase of organic carbon present in the substrate (Workneh et al., 1993). It should be noted, however, that the cover crop can sometimes increase the presence of soil-borne pathogens, especially if the latter show polyphagous and / or saprophytic edaphic behavior (Lewis and Papavizas, 1975, 1977).

In this case, special attention should be paid to the choice of cover crop species. According to Leach (1993), *Hordeum vulgare* L., *Lupinus albus* L. or *Pisum sativum* L. are useful cover crops species in order to limit the attacks by *Rhizoctonia solani* Kühn on potato crops.

3.3.2.b. Biofumigation

The containment of soil-borne pathogens by the cover crops can also be caused by biochemical mechanisms related to the release into the soil of biocidal substances. In this respect, the potentialities of certain species for the realization of the so-called biofumigation, are well documented (Kirkegaard and Mathiessen, 2004). This technique is based on the green manuring of some species belonging to the Brassicaceae family (*Brassica* spp., *Raphanus* spp.) and the subsequent release into the soil of isothiocyanates (via the hydrolysis of aliphatic and aromatic glucosinolates by the enzyme thioglucosidases-hydrolase, also called myrosinase) (Mathiessen and Kirkegaard, 2006), that induce a high mortality against a broad range of soil-borne pathogens (Bones and Rossiter, 1996; Rosa and Rodrigues, 1999; Rask et al., 2000). *In vivo* and *in vitro* experiments have demonstrated that aliphatic isothiocyanates exert limiting effects against fungi belonging to the genera *Fusarium*, *Pythium* and *Rhizoctonia*, and bacteria belonging to the genus *Ralstonia* (Manici et al., 1997, 2000). In tests conducted in Australia (Akiew and Trevorrow, 1999), the green manure of *Brassica juncea* (L.) Czern. significantly reduced the presence of *Ralstonia solanacearum* (Smith) Yabuuchi et al., with positive effects on the subsequent tobacco crop.

3.3.3. Insect Pests

Cover crops can have a key role in implementing integrated and/or biological insect pests management strategies, by enhancing the biological complexity of agroecosystems (Altieri and Schmidt, 1985; Bugg et al., 1991; Hendricks, 1995). Generally, this can derive from an improved fitness of natural enemies (biological control) or from a reduction in pests dispersal ability, reproduction or plant colonization capability (cultivation control) (Bugg, 1992).

3.3.3.a. Insect Pests Biological Control

Biological control is possible since cover crops attract beneficial arthropods, by providing them both nesting and oviposition sites (Costello and Daane, 1998; Nicholls and Altieri, 2012).

However, their effects are closely linked to the feeding behavior of beneficial arthropods. This diversification of biocenosis due to cover cropping, generally increases the presence of predators, which are polyphagous at all their ontogenetic stages. These populations, if constantly numerous, are able to prevent the sudden outbreaks of many crops pests (Bugg and Wilson, 1997; Costello and Daane, 1998).

Several species belonging to *Vicia* spp. and *Trifolium* spp. may increase the presence of predators belonging to the genera *Orius* and *Geocoris* or to Coccinellidae family (Bugg et al., 1990). *Trifolium subterraneum* L. and *T. alexandrinum* L. are able to significantly augment the presence of Geocoridae predators (Lu et al., 2000). Other cover crop species having an attractive effect on beneficial arthropods are *Fagopyrum esculentum* Moench, *Melilotus officinalis* (L.) Pallas, *Trifolium pratense* L., *T. repens* L. and *Vigna* spp (Dufour, 2000). More complex is the effect of cover crops toward those insect species limiting their predatory activity to the juvenile stages, but whose adults show glyciphagous behavior (e.g. several Neuroptera Chrysopidae). In these cases the biological activity of juveniles stages is enhanced thanks to the diversification of prey populations, while the adult fitness (longevity, fertility) is improved because of greater availability of nectar and pollen provided by the cover crop (Costello and Daane, 1998; Dufour, 2000). With reference to parasitoids, whose juvenile stages are often oligophagous or monophagous and adult stages are glyciphagous, cover cropping generally improves the adult fitness by providing a greater availability of nectar. Cover crops species best performing this function, are characterized by abundant flowering and rich nectar secretion, whose nectariferous regions are easily accessible to the adults. *Pimpinella anisum* L., *Carum carvi* L., *Anethum graveolens* L., *Petroselinum crispum* (Mill.) AW Hill, *Trifolium repens* L., *Foeniculum vulgare* Mill., *Tanacetum vulgare* L. and *Achillea millefolium* L. are very attractive with respect to several Hymenoptera Calcididae and Braconidae (Dufour, 2000). Table 1 lists some cover crops that positively influence the fitness of beneficial insects.

3.3.3.b. Insect Pests Cultivation Control

It should be noted, however, that in many cases, cover cropping could increase the insect pests pressure against the cash crop, considering the complex interactions between arthropod communities, main crop and cover crop (Lewis et al., 1997), hence the correct choice of cover crop species cannot be made without knowledge of their interaction with the most damaging pests present in a given area. Indeed, several leguminous can represent host plants with regard to insect pests such as *Agrostis*, *Peridroma* and *Heliotis* (Stadelbacher, 1981; Snodgrass et al., 1984; Dabney et al., 2001). Nevertheless, this characteristic can be positively exploited in order to implement a pest cultural control. Indeed, if properly managed, this feature can be favorably used for the production of trap crops, which represent preferential sites for pests accumulation.

Table 1. Cover crops used, predators/parasitoids attracted and controlled insect-pests (reviewed from Dufour, 2000)

Cover crop species	Predators / Parasitoids attracted	Controlled insect pests
Anethum graveolens L., *Brassica juncea* (L.) Czern., *Thymus pulegioides* L., *Melilotus* spp.	*Aphidoletes aphidimyza* Rondani	Several aphid species
Pimpinella anisum L., *Carum carvi* L., *Anethum graveolens* L., *Petroselium crispum* L., *Trifolium repens* L.	*Aphidius matricariae* Haliday	Several aphid species
Trifolium alexandrinum L., *T. subterraneum* L.	*Geocoris* spp.	Several insects and mites
Carum carvi L., *Anethum graveolens* L., *Petroselium crispum* L., *Trifolium repens* L., *Achillea millefolium* L., *Vicia villosa* Roth., *Fagopyrum esculentum* Moench, *Vigna unguiculata* (L.) Walpers	Hymenoptera Braconidae	*Spodoptera* spp., *Pieris* spp., *Cydia pomonella* L., *Lymantria dispar* L., *Ostrinia nubilalis* Hübner
Solidago spp, *Achillea millefolium* L., *Medicago sativa* L.	Heteroptera Nabidae	Several aphid, thrips and leafhoppers
Fagopyrum esculentum Moench, *Carum carvi* L.,*Coreopsis* spp., *Solidago* spp.	Neuroptera Crysopidae	Several thrips, leafhoppers and mites species
Trifolium repens L.	Coleoptera Carabidae	Several slugs species. Among the insects, *Delia radicum* L., *Lymantria dispar* L., *Malacosoma* spp.
Foeniculum vulgare Mill., *Anethum graveolens* L., *Solidago* spp., *Coreopsis* spp., *Trifolium incarnatum* L., *Vicia villosa* Roth., *Fagopyrum esculentum* Moench, *Secale cereale* L., *Cannabis sativa* L., *Sesbania exaltata* (Raf.) Cory	*Hippodamia* spp.	Several aphids, mealybugs and mites species

Cover crop species	Predators / Parasitoids attracted	Controlled insect pests
Foeniculum vulgare Mill., *Anethum graveolens* L., *Solidago* spp., *Coreopsis* spp., *Achillea millefolium* L.	*Cryptolaemus montrouzieri* Mulsant	Several mealybugs species
Coriandrum sativum L.	Diptera Syrphidae	Several aphids species
Phacelia spp.	Diptera Tachinidae	*Agrostis* spp., *Spodoptera* spp., *Malacosoma* spp., *Trichoplusia* spp., *Lymantria dispar* L.; *Popillia japonica* Newman, *Phyllophaga* spp., Coreidae, *Acrosternum hilare* Say
Anethum graveolens L., *Pimpinella anisum* L., *Carum carvi* L., *Vicia villosa* Roth., *Mentha viridis* Auct., *Fagopyrum esculentum* Moench, *Achillea millefolium* L., *Trifolium repens* L., *Vigna unguiculata* (L.) Walpers, *Foeniculum vulgare* Mill., *Anthriscus cerefolium* (L.) Hoffm.	Hymenoptera Trichogrammatidae	*Choristoneura* spp., *Helicoverpa armigera* Hübner, *H. zea* Boddie, *Manduca quinquemaculata* Haworth, *Ostrinia* spp., *Cydia pomonella* L.
Anethum graveolens L., *Foeniculum vulgare* Mill., *Achillea millefolium* L., *Coreopsis* spp.	*Encarsia formosa* Gahan	*Trialeurodes vaporariorum* Westwood, *Bemisia tabaci* Gennadius

In these cases, the suppression of trap crops through mechanical destruction and burial of biomass, can promote a high mortality of pest populations, helping to diminish their pressure toward the cash crop (Kuepper, 2003; Trdan et al., 2005).

3.3.4. Plant-Parasitic Nematodes

Plant-parasitic nematodes are the primary cause for severe yield losses in a multitude of cash crops cultivated worldwide. They may reduce crop yields through compromising root functionality, directly vectoring virus diseases, or promoting plant colonization by soil-borne bacteria and fungi through root lesions (Hooks et al., 2010). Hence, the possibility of controlling plant parasitic nematodes through cover cropping is a highly topical research theme, especially for its potential impact in cropping systems sustainability (Reddy et al., 1986; Yeates, 1987; Rodriguez-Kàbana et al., 1988). Nonetheless, current knowledge does not allow the implementation of strategies to successfully control nematodes in all agroecosystems typologies. This is attributed to the marked biological plasticity of this zoological group, resulting in the differentiation of races, biotypes and pathotypes, especially within the more common species, so often changing the relationships between cover crops and nematode species (Caswell and Roberts, 1987). Moreover, researches conducted to date have mainly been aimed at studying the effects of cover crops on single or very few nematode species, and this while having to tackle the broad systematic diversification of nematode populations inhabiting soil agroecosystems (Westerdahl et al., 1998). The control of plant-parasitic nematodes by cover cropping is based on the assumption that they are obliged parasitic life forms, so the presence of host plants is a necessary condition in order to complete their life cycle (Freckman and Caswell, 1985). Another element is the need for a match between sexes that, due to the low mobility of nematodes within the soil, offers other possibilities to interfere in their ontogenetic cycle. Cover crops can affect nematode populations consistency either through direct or indirect effects.

3.3.4.a. Direct Nematodes Control

Cover crops directly affect the feeding and reproduction behavior of nematodes by some mechanisms which can provide (i) the host plant-nematode incompatibility, (ii) the release, by non-host plants, of root exudates promoting nematode egg hatching, (iii) the blocking of sexual maturation of those nematodes penetrating into the roots (trap crops), (iv) the release of root exudates disturbing the search for the host plants or the search between sexes, (v) the sustaining of low reproductive rates that are insufficient to compensate the mortality of nematode communities or (vi) releasing in the soil nematicidal compounds (biofumigation) (Westerdahl et al., 1998; Thorup-Kristensen et al., 2003).

Another type of direct control is represented by the use of trap crops. Such mechanisms can occur separately or in combination, resulting in a lower presence of plant-parasitic nematodes within the soil. A well documented case in the literature concerns the control of the nematode *Heterodera schactii* Schmidt through the cultivation of *Raphanus sativus* L. and *Sinapis alba* L.

(Muller, 1999), whose roots are colonized by the mobile forms of the nematode, which, however, fail to sexually mature due to the glucosinolate - myrosinase system (Thorup-Kristensen et al., 2003). This reduces the number of females in the following generation, causing a progressive decline in nematode population and, therefore, reduced attacks against the succeeding cash crop (Mathiessen and Kirkegaard, 2006). With reference to biofumigation, some Brassicaceae [*Brassica nigra* (L.) WDJ Koch, *B. juncea* (L.) Czern., etc.] with a high content of propenyl-glucosinolate, show significant nematicidal effects against *Meloidogyne javanica* (Treub) Chitwood (McLeod and Steele, 1999) and *Globodera pallida* (Stone) Behrens (Lord et al., 2011). Similarly, the green manure of *Tagetes* spp., are able to significantly reduce the presence of *Pratylenchus vulnus* Allen and Jensen, *P. penetrans* (Cobb) Filipjev Stekhoven and Schuurmans, *P. brachyurus* (Godfrey) Filipjev and Schuurmans Stekhoven and *Paratrichodorus minor* (Colbran) Siddiqi (MacDonald and Mai, 1963; Reddy et al., 1986; McKenry, 1991). Other species with significant nematicidal effects are *Sorghum bicolor* (L.) ssp. *drummondii* (Nees ex Steud.) de Wet and Harlan, *Indigofera hirsuta* Harvey, *Crotalaria spectabilis* Roth., *C. juncea* L. and *Mucuna deeringiana* (Bort) Merr (Araya and Caswell-Chen, 1994a, b).

3.3.4.b. Indirect Nematodes Control

Indirect nematodes control is achieved mainly through the organic matter supply to the soil, which favors the enhancing of soil biodiversity and microbial biomass, so increasing the presence of nematodes' natural enemies, including endophytic bacteria and nematode-trapping fungi (Hooks et al., 2010; Nair and Ngouajio, 2012). In a three-year study, Buyer et al. (2010) demonstrated that the integration of both rye and vetch cover crop into crop rotations, signiticatly increased soil microbial biomass.

Moreover, it was found that the green manure of alfalfa and barley increases the presence of the nematophagous fungus *Drechmeria coniospora* (Drechsler) Gams and Jansson (Van den Boogert et al., 1994) and that the green manure of phacelia, barley and pea causes an increase of soil bacteria and fungi, resulting in a reduction of nematode populations (Brzeski et al. 1993). Wang et al. (2002) demonstrated that *Tagetes erecta* L. increased nematode-trapping fungal population densities one month after soil incorporation.

However, the Authors pointed out that the enhancement of nematode-trapping fungi was a short-term effect and no longer detectable in the ordinary field conditions.

4. COVER CROP SPECIES

The most common cover crop species belong to three botanical families, i.e. Fabaceae, Poaceae and Brassicaceae. They have marked differences in their biological behavior, as a reflection of their different morphological, ecological and physiological traits. It must be noted that, even within the same species, different cultivars can prove significantly different in their bio-agronomical characteristics (in terms of earliness, adaptability, biomass yield), both among diverse locations and within a same locations, so the following paragraphs should only be considered general guidelines focusing on the main differences among cover crop species.

4.1. Fabaceae

Leguminous cover crops owe their popularity most of all to their ability to fix atmospheric nitrogen via root rhizobia, a feature which is of pivotal importance when increasing soil chemical fertility is an expected benefit from cover cropping. Particularly important legumes are some clovers (*Trifolium* spp.) and medics (*Medicago* spp.) characterized by annual life cycle, self-reseeding ability (especially clovers) and by significant production of hard seeds. This last mechanism have a functional role in the sward persistence over time, especially in those environments with uneven rainfall distribution, especially during the warm period. Other popular cover crop legumes belong to the genera *Pisum* and *Vicia.* Among clovers, an important place is taken by the subterranean clovers species complex, as well as by *Trifolium incarnatum* L., *T. alexandrinum* L., *T. repens* L., *T. hirtum* All. and *T. fragiferum* L. The subterranean clover complex includes *Trifolium subterraneum* L., *T. brachycalycinum* Katzn. et Morley and *T. yanninicum* Katzn. et Morley; a fourth species (*T. israeliticum* Zoh. and Katzn.) was also identified, but it is considered minor (Zohary and Katznelson, 1958; Talamucci, 2002). They are native to the Mediterranean Basin and coastal areas of Western Europe (Katznelson, 1974; Mauromicale et al., 1997), performing their life cycle during autumn-spring period. They have short stature (15-30 cm), shallow roots, tomentose stems and leaves, these last with whitish spots on their laminae, useful for cultivar identification. At seeds ripening, the numerous flowers tend to stratify themselves within and above the soil, thanks to the species' geocarpy, making them excellent at self-reseeding (Pecetti and Piano, 2002; Mauromicale et al., 1997, 2010).

The fruit is composed of 3-5 pods, each generally with a single seed, and from the residues of the perianth, all enclosed in a proliferation of sterile calyxes. At seed maturation, the percentage of hard seeds varies from 65 to 90%, in relation to the cultivars. *Trifolium subterraneum* L. prefers well-drained soil, relatively low pH (from neutral to sub-acid), while *T. brachycalycinum* Katzn. et Morley is adaptable to various soil kinds, especially those of medium texture, with sub-acid, neutral or alkaline pH. *T. yanninicum* Katzn. et Morley prefers loamy and moist soils, with an acidic, sub-acid or neutral pH. In recent decades the subterranean clovers were the subject of intensive breeding programs, leading to the release of over 50 different cultivars, differing from each other for agronomic characteristics and adaptability to different growing environments. In Mediterranean areas, good results have been achieved by the Spanish cultivars 'Orellana' (early) and 'Areces' (medium-early) belonging *Trifolium subterraneum* L. species, and 'Valmoreno' (semi-late) and 'Gaitan' (late) belonging to *Trifolium brachycalycinum* Katzn. et Morley. Subterranean clovers are particularly suitable for cover cropping, but the economic and environmental problems related to seed production are gradually reducing the attractiveness of these species, thus prompting the search for alternative species.

Trifolium incarnatum L. (crimson clover) is an annual species with shallow roots, poorly branched stems, hairy and heart-shaped leaves. The inflorescence is a terminal, cylindrical capitulum with dark-red flowers. The seeds are polished, oval, yellowish. The 1000 seeds weight ranges between 3.2 - 3.6 g. It is fairly tolerant to low temperatures, but undergoes strong thinning during winter if grown on calcareous clay soils, where it is subject to being uprooted by frost and/or attacked by *Sclerotinia trifoliorum* Erikss. In order to better manage the living mulch, 3 to 5 cuttings per year are required, which should be performed when the plant reaches a height ranging from 8 to 12 cm. To enable a good self-reseeding, cuttings must be discontinued near flowering.

Trifolium alexandrinum L. (berseem clover) is one of most interesting fodder crops, particularly widespread in North Africa (especially in Egypt), Middle East and India. Its cultivation has recently attracted growing interest in North and South America, South Africa, Australia and Southern Europe. It is an annual plant with very deep tap-roots, branched and erect stems, oblong and broadly elliptical leaflets, and yellow or white flowers, grouped in elliptical inflorescences. In mild winter climates, berseem clover grows quickly during winter, blooming in spring, but earlier than other clover species. Due to the good tolerance to waterlogging, it can be used to remove the excessive moisture present in the soil.

The dead mulch is rich in organic nitrogen and, under optimal conditions, its biomass can add more than 350 kg ha^{-1} year^{-1} of nitrogen to the soil (Williams et al., 1990).

Trifolium repens L., commonly known as white clover, is a particularly common species in the natural pastures of the Euro-Asian area, thanks to its remarkable environmental adaptability. It is a perennial species with creeping stems and is able to develop adventitious roots from stem nodes. The leaves have slightly oval leaflets, denticulated throughout the margin and often with a light green spot. The flowers are white with light pink hues, gathered in a large numbers of capitula, carried by long, erect stalks, reaching a higher level than that of the leaves. The seeds are very small and vary from yellow to reddish. The white clover is well adapted to temperate and humid climates, prefers sandy soils, with a good content of calcium. It is capable of good biomass yields even in loamy soils, even if not particularly rich in minerals.

Trifolium hirtum All. (rose clover) has semi-erect, branched, slightly pubescent stems, hairy and egg-shaped leaflets with reddish-white spots, broad stipules, light-pink flowers, solitary and rounded. It can thrive in a wide range of soil types, provided that they are well drained. It has a good self-reseeding ability and, thanks to the high seeds yield, shows good competitive ability against weeds. Among the cultivars available, 'Hykon' (early) is well suited for cover cropping purposes

Trifolium fragiferum L. (strawberry clover) has branched, creeping stem, broadly elliptical, slightly ovate leaves, pink terminal flowers, abundant flowering. Each runner allows the plant to resettle annually, even in neighboring areas. It tolerates very wet, alkaline and saline soils, endures frequent mowing, can effectively control soil erosion and is resistant to a wide range of herbicides. However, it should be noted that it competes with the main crop for water. The cultivar 'Salina' adapts well to the Mediterranean environment. Among medics, *Medicago scutellata* (L.) Miller and *M. polymorpha* L. (syn. *M. hispida* Gaertner) have a certain importance. The former species (also called snail medic), originates from the Mediterranean Basin, is an annual species with erect or semi-erect growth habit, and rather tall stature (70-130 cm). The leaves are obovate, with a slight marginal indentation; the upper lamina is glabrous while the lesser is tomentose. It produces lateral inflorescences consisting of racemes with a few (1-5) yellow flowers. Legumes, hemispherical, with 4-6 counterclockwise turns and spineless, when ripe are detached from the plant and, given their size (1.2 - 2 cm in diameter), are easily ingested by livestock during grazing. 75-80% of seeds show hardseededness.

It is well adapted to clay soils, with high limestone content and a pH varying from 6.5 and 9. It is very sensitive to low temperatures. In order to reach its full yield potential, the species requires at least 450-500 mm annual rainfall, well distributed during the vegetative stage. Some commercially available cultivars of Australian origin are 'Kelson', 'Saba', 'Sair' and 'Robinson'. It is able to compete effectively with weeds (Occhipinti et al., 2007).

Medicago polymorpha L. (burr medic) shows semi-erect or prostrate stems, ovate and heart-shaped leaves, slightly indented, rather small stipules, and small yellow flowers, twisted pods with thorns in the apex. Each pod contains from 3 to 11 seeds. Burr medic is a good cover crop species thanks to its high growth rate, good self-reseeding ability and wide adaptability to soil pH.

4.2. Poaceae

The species belonging to the Poaceae family (grasses) are characterized by highly branched and developed root systems, hence they are ideal cover crop species for reducing soil erosion and increasing water infiltration. Moreover, they have both high growth rates and N-demands, and can remove from the soil as much as up to 2 kg ha^{-1} d^{-1} of N (Ingels et al., 1998), hence are often utilized as catch crops, in order to reduce N leaching (Lu et al., 2000). To this end, excellent catch crops are *Avena sativa* L., *Hordeum vulgare* L., *Secale cereale* L., *Lolium multiflorum* Lam., *Bromus hordeaceus* L., *Festuca arundinacea* Schreb. and *Dactylis glomerata* L.

Avena sativa (oat) is an annual, winter-hardy species, with expanded and deep root system. The culms are robust, with blue-green leaves and a wide blade and highly developed ligule, but without auricles. The inflorescence is a panicle, whose branches bear spikelets each with two flowers. At maturity, the caryopsis is enclosed within the integuments. This species is often used in mixtures, sometimes in pure culture. It tolerates very moist and clay soils, moderately fertile and even with a pH below 4.5.

Hordeum vulgare (barley) is an annual, winter-hardy species. The leaves are different from those of other cereals for the large auricles and a very small ligule. The inflorescence is a compact spike, whose rachis is made up of 20-30 articles. On each node are inserted three spikelets, each with a single flower with two lateral sterile glumes. The husks are well developed and adhere closely to the caryopsis.

The inferior husk almost always ends with a long and strong loin. As a result of its outstanding tillering ability, barley also produces high amounts of aboveground biomass, and therefore competes effectively with weeds. More than rye and oat, it tolerates soil salinity and water scarcity, but is sensitive to excess soil moisture.

Secale cereale (rye) is an annual species with thin and flexible culms reaching up to 2 m height. The leaf laminae are blue-green colored and relatively short. The spike is long and narrow. On each rachis node there is an inserted spikelet, which normally has three flowers. It thrives well even on sandy, poor soils. Its dead mulch has a higher persistence than that of other cereals and has a strong allelopathic effects against some weeds and plant-parasitic nematodes (Ercoli et al., 2007; Zasada et al., 2007). 'Merced' is a well suited cultivar for cover cropping purposes, as it is characterized by a good adaptability to a wide range of soil conditions, and has elevated growth rate during winter.

Lolium multiflorum (Italian ryegrass) has thick culm, pale yellow at its base, dark green leaves, prominent auricles, inflorescence (panicle) upright and relatively narrow. It has great adaptability and good self-reseeding ability. It is often found within the natural flora. Among grasses, it is a species that flourishes well and grows more quickly, so it competes effectively with weeds.

Bromus hordeaceus is an annual or biannual grass species, having erect culms, slightly pubescent leaves, small, whitish and membranous auricles, compact inflorescence (panicle). It thrives on various soil types and under a wide range of temperatures. It responds well in reduced-tillage cropping systems, as for example, the vineyards. It grows slowly and does not compete with the main crop and has a good self-reseeding ability. It can be used either in pure culture or in mixtures with low-sized species.

Festuca arundinacea (tall fescue) has an erect, robust culm, thin and resistant leaves with rough edges, a narrow inflorescence (panicle) highly branched. It is a fast growing perennial species, so it can be used to reduce the excess of vigor of many crops (e.g. grapevine). It tolerates the summer drought well and likewise low winter temperatures and waterlogging. In the US, the cultivar 'Fawn', albeit requiring several mowings per year, is commonly used for cover cropping in vineyards (Ingels et al., 1998).

Dactylis glomerata (orchard grass) is a bushy plant, with green-glaucous leaves, without auricles, but with a long white ligule. It has marked persistence (5-8 years), although lower than that of *F. arundinacea*. Because of its fairly good vigor and ability to compete, may be used when it is necessary to reduce the vigor of the cash crop (Wolpert et al., 1993).

It tolerates both cold winters and long periods of summer drought and thrives in many soil types, not overly long subject to waterlogging. It reaches its maximum height in the spring. The cultivar 'Barber' is recommended for cover cropping in vineyards (Ingels et al., 1998).

4.3. Brassicaceae

Among the Brassicaceae, the most popular species are *Brassica hirta* Moench (syn. *Sinapis alba* L.) (white mustard), *B. juncea* (L.) Czern. (Indian mustard), *B. napus* L. (rapeseed), *B. nigra* (L.) W.D.J. Koch (black mustard) and *B. rapa* L. (turnip) (syn. *Brassica campestris* L.), characterized by branched and erect stems, leaves of different shape and size in relation to their position on the plant, yellow or whitish flowers with four petals. The fruit is a capsule, often carrying hardseeds. These species are considered suitable for cover cropping purposes in Mediterranean climate areas because they:

- Settle and complete the biological cycle during the autumn-spring period, when their water needs can be satisfied by rain, hence they can be used in those cropping systems that apply localized irrigation;
- Have a deep root system, which effectively contributes to increasing the aeration and water infiltration in clay soils;
- Are able to absorb substantial amounts of nitrates, subtracting them from leaching;
- Produce biomass with a low C/N ratio and, therefore, easily mineralizable after burial.

Many of the *Brassicaceae* are characterized by content of glucosinolates, a class of glycosidic compounds that contain sulphur and have defensive function against natural enemies. When tissues are injured, glucosinolates are hydrolyzed and may become isothiocyanates, volatile sulfur-based allelotoxins. Degradation of *Brassicaceae* cover crops residues, following their incorporation in soil, can be exploited for soil biofumigation, considering effectiveness of isothiocyanates against a wide range of soil-borne plant pathogens and nematodes. The use of *Brassicaceae*, especially Indian mustard and black mustard, for their high content of glucosinolates, is being studied to replace or reduce resorting to chemical fumigation (Trott et al., 2012).

The allelopathic properties of *Brassicaceae* may also be exploited against weeds.

Some authors have shown, in field conditions, that soil incorporation of residues of rapeseed, canola and yellow and white mustard lessens establishment and biomass growth of weeds (Alcántara et al., 2011). However, we should also note that *Brassicaceae*, for the same reason that they inhibit weeds, may have detrimental effects on the establishment, growth and yields of crops in succession.

CONCLUSION

Cover cropping may have a central role in improving the efficiency of farming systems, it can improve soil fertility, increase the ecological stability of the agroecosystems and yield capacity of cash crops, reducing, at the same time, the need for external inputs (Campiglia, 1999). It should be noted, however, that the further expansion of cover crops in agricultural systems cannot be separated from the improvement of some agronomic and genetic aspects of cover cropping. On the first aspect, greater attention should be paid to the management of the living mulches, especially in order to reduce management costs and mitigate the competition of the cover crops toward the main crop. Another aspect is the development of improved mixtures, able to enhance the ecological complementarity between associated species. Concerning genetic aspects, it must be noted that currently cover cropping, in most cases, is performed by utilizing cultivars constituted for conventional uses (grain or forage). Breeding for specific genotypes appears, therefore, clearly desirable (Mauromicale et al., 2010) and those cropping systems characterized by low efficiency in utilizing native resources would be particularly useful. The goals of genetic improvement should stem from the analysis of the specific attitudes of the utilized species. As regards the legume species, it would appear advantageous to increase their self-reseeding ability (especially in the annual medics) and tolerance to high shading levels, which are typical of modern orchards (Campiglia and Caporali, 1996; Mauromicale et al., 2010; Mauro et al., 2011). Regarding grasses, a primary target of breeding might be the exaltation of the speed of settlement during the autumn and their earliness in completing their cycle could help both to improve their action against soil erosion and minimize their edaphic competition against consociated or succeeding crops. In the Brassicaceae, the enhancement of biocidal activity could allow the reduction of the use of soil fumigants, with a significant improvement of the ecological profile of many intensive cropping systems. It would also prove beneficial to mitigate their potential phytotoxic effects towards cash crops.

At present, these possible innovations appear one of most rational way to greatly improve the performance of cover crops and to facilitate their more widespread integration, even in highly specialized cropping systems.

REFERENCES

Abdul-Baki, A., Teasdale, J. R. and Korcak, R. F. (1997) Nitrogen requirements of fresh-market tomatoes on hairy vetch and black polyethylene mulch. *HortScience* 32, 217–221.

Adler, S. J. (1995) Orchardists can reap benefits from cover crops. *Good fruit grower* 46 (5), 39-40.

Akiew, S. and Trevorrow, P. (1999) Biofumigation of bacterial wilt of tobacco. In: Magarey, R. C. (Ed.), *Proceedings 1st Australasian Soilborne Disease Symposium.* Bureau of Sugar Experiment Stations, Brisbane (Australia), 207-208.

Alcántara, C., Pujadas, A. and Saavedra, M. (2011) Management of cruciferous cover crops by mowing for soil and water conservation in southern Spain. *Agricultural Water Management* 98 (6): 1071–1080.

Altieri, M. A. and Schmidt, L. L. (1985) Cover crop manipulation in Northern California orchards and vineyards: Effects on arthropod communities. *Biological Agriculture and Horticulture* 3: 1-24.

Araya, M. and Caswell-Chen, E. P. (1994a) Penetration of *Crotalaria juncea, Dolichos lablab* and *Sesamum indicum* roots by *Meloidogyne javanica. Journal of Nematology* 26: 238-240.

Araya, M. and Caswell-Chen, E. P. (1994b) Host status of *Crotalaria juncea, Sesamum indicum, Dolichos lablab* and *Elymus glaucus* to *Meloidogyne javanica. Journal of Nematology* 26: 492-497.

Ashworth, S. and Harrison, H. (1983) Evaluation of mulches for use in the home garden. *HortScience,* 18: 180-182.

Batey, T. (2009) Soil compaction and soil management. *Soil Use and Management,* 25: 335-345.

Baumeister, J. (1964) Zur Wasserversorgung der Boden in grassgemullchter Obstalngen. *Der Erwerbsotsbau* 4: 73-76.

Bergkvist, G., Stenberg, M., Wetterlind, J., Bath, B., and Elfstrand, S. (2011) Clover cover crops under-sown in winter wheat increase yield of subsequent spring barley-effect of N dose and companion grass. *Field Crops Research* 120 (2): 292–298.

Bialy, Z., Oleszek, W., Lewis, J., and Fenwick, G. R. (1990) Allelopathic potential of glucosinolates (mustard oil glycosides) and their degradation products against wheat. *Plant and Soil* 129: 277-281.

Bonari, E., Mazzoncini, M., Menini, S., and Moschetta, D. (1997) Effetti delle "cover crops" nella monosuccessione del mais (*Zea mays* L.) da granella in terreni limosi. *Agricoltura Ricerca* 169: 39-52.

Bond, J. J. and Willis, W. O. (1969) Soil water evaporation: surface residue rate and placement effects. *Proceedings of the Soil Science Society of America,* 33: 445-448.

Bones, A. M. and Rossiter, J. T. (1996) The myrosinase-glucosinolate system, its organisation and biochemistry. *Physiologia Plantarum* 97: 194–208.

Bradow, M. and Connick, W. J. (1990) Volatile seed germination inhibitors from plant residues. *Journal of Chemical Ecology* 16: 645-666.

Breland, T. A. (1995) Green manuring with clover and ryegrass catch crops undersown in spring wheat effects on soil structure. *Soil Use Management,* 11: 163-167.

Brzeski, M. W., Smolinska, U., Szczech, M., Paul, M., and Ostrzycka, J. (1993) Short-term effect of green manuring on soil-inhabiting nematodes and microorganisms. *Nematologia Mediterranea* 21: 169-176.

Bugg, R. L. (1992) Using cover crops to manage arthopods on truck farm. *HortScience* 27: 741-745.

Bugg, R. L., Wackers, F. L., Brunson, K. E., Phatak, S. C., and Dutcher, J. D. (1990) Tarnished plant bug, *Lygus lineolaris* (Palisot de Beauvois) (Hemiptera: Miridae), on selected cool season leguminous cover crops: abundance, survival and performance. *Journal of Entomological Science* 25: 463-474.

Bugg, R. L., Wackers, F. L., Brunson, K. E., Dutcher, J. D., and Phatak, S. C. (1991) Cool-season cover crops relay intercropped with cantaloupe: influence on a generalist predator, *Geocoris punctipes* (Hemiptera: Lygaeidae). *Journal of Economic Entomology* 84: 408-416.

Bugg, R. L. and Wilson, L. T. (1997) *Ammi visnaga* (L.) Lamarck (Apiaceae): Associated beneficial insects and implications for biological control with emphasis on the bell-pepper ecosystem. *Biological Agriculture and Horticulture* 6: 241–268.

Burket, J. Z., Hemphill, D. D. and Dick, R. P. (1997) Winter cover crops and nitrogen management in sweet corn and broccoli rotations. *HortScience* 32: 664-668.

Buyer, J. S., Teasdale, J. R., Roberts, D. P., Zasada, I. A., and Maul, J. E. (2010) Factors affecting soil microbial community structure in tomato cropping system. *Soil Biology and Biochemistry* 42: 831-841.

Campiglia, E. (1999) Colture di copertura utilizzate in agroecosistema mediterranei. Nota 1: modificazioni dell'ambiente colturale. *Rivista di Agronomia,* 33: 90-103.

Campiglia, E. and Caporali, F. (1996) Valutazione del regime e modalità di defoliazione del trifoglio sotterraneo (*Trifolium subterraneum* L.) per l'impiego come "cover crop" in frutteti specializzati. *Rivista di Agronomia,* 30 (1): 86-91.

Caporali, F. and Campiglia, E. (1994) Innovazioni nella ricerca di sistemi colturali a bassi input: l'impiego del trifoglio sotterraneo (*Trifolium subterraneum* L.) come "living mulch" e sovescio nella rotazione frumento (*Triticum aestivum* L.) - girasole (*Helianthus annuus* L.) Nota II. *Rivista di Agronomia* 28 (1): 50-56.

Caporali, F. and Campiglia, E. (2001) Leguminose come cover crop in ambiente mediterraneo. In: Calderini Edagricole (Ed.), *Leguminose e agricoltura sostenibile. Specie da granella e cover crops.* Edizioni Agricole Calderini, Bologna (Italy), 149–178.

Caporali, F., Campiglia, E. and Paolini, R. (1990) Annual self-regenerating legumes for low input intensive cropping system. *Proceedings of the 1st International Symposium on "Natural Resources Management for a Sustainable Agriculture"* February 6-10, New Dehli, India, 3-14.

Campiglia, E., Mancinelli, R. and Radicetti, E. (2011) Influence of no-tillage and organic mulching on tomato (*Solanum lycopersicum* L.) production and nitrogen use in the Mediterranean environment of central Italy. *Scientia Horticulturae* 130: 588-598.

Campiglia, E., Radicetti, E. and Mancinelli, R. (2012) Weed control strategies and yield response in a pepper crop (*Capsicum annuum* L.) mulched with hairy vetch (*Vicia villosa* Roth.) and oat (*Avena sativa* L.) residues. *Crop Protection* 33: 65-73.

Caswell, E. P. and Roberts, P. A. (1987) Nematode population genetics. In: Veech, J. A., Dickson, D. W. (Eds.), *Vistas on nematology.* Hyattsville, MD: Society of Nematology.

Cherr, C. M., Scholberg, J. M. S. and McSorley, R. (2006) Green manure approaches to crop production: a synthesis. *Agronomy Journal,* 98: 302-319.

Clark, A. J., Decker, A. M. and Meisinger, J. J. (1994) Seeding rate and kill date hairy vetch-cereal rye cover crop mixtures for corn production. *Agronomy Journal* 86: 1065-1070.

Cline, G. R. and Silvernail, A. F. (2001) Residual nitrogen and kill date effects on winter cover crop growth and nitrogen content in a vegetable production system. *HortTechnology* 11: 219–225.

Columella. De Re Rustica. *Loeb Classical Library,* 1941.

Costello, M. J. and Daane, K. M. (1998) Arthropods. In: Ingels, C. A., Bugg, R. L., McGourty, G. T., Christensen, L. P. (Eds.), *Cover cropping in vineyards. A grower's handbook.* University of California, Agriculture and Natural Resources Communication Services, Oakland (US), 93-106.

Cotching, W. E., Allbrook, R. F. and Gibbs, H. S. (1979) Influence of maize cropping on the structure of two soils in the Waikato district, New Zealand. *New Zealand Journal of Agricultural Research,* 22: 431-438.

Dabney, S. M., Delgado, J. A. and Reeves, D. W. (2001) Using winter cover crops to improve soil and water quality. *Communications in Soil Science and Plant Analysis,* 32 (7-8): 1221-1250.

De Baets, S., Poesen, J., Meersmans, J., and Serlet, L. (2011) Cover crops and their erosion-reducing effects during concentrated flow erosion. *Catena* 85 (3): 237–244.

Decker, A. M., Clark, A. J., Meisinger, J. J., Mulford, F. R., and McIntosh, M. S. (1994) Legume cover crop contributions to no-tillage corn production. *Agronomy Journal* 86: 126–135.

Den Hollander, N. G., Bastiaans, L. and Kropff, M. J. (2007a) Clover as cover crop for weed suppression in an intercropping design I. Characteristics of several clover species. *European Journal of Agronomy* 26: 92-103.

Den Hollander, N. G., Bastiaans, L. and Kropff, M. J. (2007b) Clover as cover crop for weed suppression in an intercropping design II. Competitive ability of several clover species. *European Journal of Agronomy* 26: 104-112.

Dufour, R. (2000) Farmscaping to enhance biological control. *National Center for Appropriate Technology,* Fayetteville, Arkansas (US). ATTRA publication, 40 pp.

Enache, A. J. and Ilnicki, R. D. (1990) Weed control by subterranean clover (*Trifolium subterraneum*) used as a living mulch. *Weed Technology* 4: 534–538.

Ercoli, L., Masoni, A., Pampana, S., and Arduini, I. (2007) Allelopathic effects of rye, brown mustard and hairy vetch on redroot pigweed, common lambsquarter and knotweed. *Allelopathy Journal* 19: 249-256.

Erisman, J. W., Bleeker, A., Galloway, J., and Sutton, M. S. (2007) Reduced nitrogen in ecology and environment. *Environmental Pollution* 150: 140-149.

Evans, P. S. (1977) Comparative root morphology of some pasture grasses and clovers. *New Zealand Journal of Science* 20: 331-335.

Fatima, T., Teasdale, J. R., Bunce, J., and Mattoo, A. K. (2012) Tomato response to legume cover crop and nitrogen: differing enhancement patterns of fruit yield, photosynthesis and gene expression. *Functional Plant Biology* 39 (3): 246-254.

Frankenberger, W. T. and Abdelmagid, H. M. (1985) Kinetic parameters of nitrogen mineralization rates of leguminous crops incorporated into soil. *Plant and Soil* 87: 257-271.

Freckman, D. W. and Caswell, E. P. (1985) The ecology of nematodes in agroecosystems. *Annual Review of Phytopathology* 23: 275-296.

Gardner, W. K., Parbery, K. G. and Barber, D. A. (1981) Proteoid root morphology and function in *Lupinus albus L. Plant and Soil* 60: 143–147.

Gardner, W. K., Barber, D. A. and Parbery, K. G. (1983) The acquisition of phosphorus by *Lupinus albus* L. III. The probable mechanism by which phosphorus movement in the soil/root interface is enhanced. *Plant and Soil* 70: 107–124.

Glenn, D. M. and Welker, W. V. (1989) Orchard soil management system influence rainfall infiltration. *Journal of the American Society for Horticulture Science* 114: 10-14.

Griffin, T., Liebman, M. and Jemison, J. Jr. (2000) Cover crops for sweet corn production in a short-season environment. *Agronomy Journal* 92: 144–151.

Hargrove, W. L. (1986) Winter legumes as a nitrogen source for no-till grain sorghum. *Agronomy Journal* 78: 70-74.

Hendricks, L. C. (1995) Almond growers reduce pesticide use in Merced County field trial. *California Agriculture* 49: 5-10.

Hirschfelt, D. J. (1998) Soil fertility and vine nutrition. In: Ingels, C. A., Bugg, R. L., McGourty, G. T., Christensen, L. P. (Eds.), *Cover cropping in vineyards. A grower's handbook.* University of California, Agriculture and Natural Resources Communication Services, Oakland (US), 61-68.

Hoffman, M. L., Weston, L. A., Snyder, J. C., and Regnier, E. E. (1996) Allelophatic influence of germinating seeds and seedlings of cover crops on weed species. *Weed Science* 44: 579-584.

Holderbaum, J. F., Decker, A. M., Meisinger, J. J., Mulford, F. R., and Vough, L. R. (1990) Fall-seeded legume cover crops for no-tillage corn in the humid East. *Agronomy Journal* 82: 117-124.

Hooks, C. R. R., Wang, K. H., Ploeg, A., and McSorley, R. (2010) Using marigold (*Tagetes* spp.) as a cover crop to protect crops from plant-parasitic nematodes. *Applied Soil Ecology* 46: 307-320.

Huntington, T. G., Grove, J. H. and Frye, W. W. (1985) Release and recovery of nitrogen from winter annual cover crops in no-till corn production. *Communications in Soil Science and Plant Analysis* 16: 193-211.

Hutchinson, C. M. and McGiffen, M. E. (2000) Cowpea cover crop mulch for weed control in desert pepper production. *HortScience* 35: 196-198.

Ierna, A., Mauro, R. P. and Mauromicale, G. (2012) Improved yield and nutrient efficiency in two globe artichoke genotypes by balancing nitrogen and phosphorus supply. *Agronomy for Sustainable Development* 32: 773-780.

Inderjit, Duke S. O. (2003) Ecophysiological aspects of allelopathy. *Planta* 217: 529-539.

Ingels, C. A. and Klonsky, K. M. (1998) Historical and current uses. In: Ingels, C. A., Bugg, R. L., McGourty, G. T., Christensen, L. P. (Eds.), *Cover cropping in vineyards. A grower's handbook.* University of California, Agriculture and Natural Resources Communication Services, Oakland (US), 3-7.

Ingels, C. A., Bugg, R. L. and Thomas, F. L. (1998) Cover crop species and descriptions. In: Ingels, C. A., Bugg, R. L., McGourty, G. T., Christensen, L. P. (Eds.), *Cover cropping in vineyards. A grower's handbook.* University of California, Agriculture and Natural Resources Communication Services, Oakland (US), 8-42.

Jansson, S. L. and Persson, J. (1982) Mineralization and immobilization of soil nitrogen. In: Stevenson, F. J. (Ed.), *Nitrogen in agricultural soils.* American Society of Agronomy, Madison (US), 229-252.

Karpenstein-Machan, M. and Stuelpnagel, R. (2000) Biomass yield and nitrogen fixation of legumes monocropped and intercropped with rye and rotation effects on a subsequent maize crop. *Plant and Soil* 218: 215–232.

Katznelson, J. (1974) Biological flora of Israel. 5. The subterranean clovers of *Trifolium* subsect. *Calycomorphum* Katzn. *Trifolium subterraneum* (sensu lato). *Israeli Journal of Botany* 23: 69-108.

Kirkegaard, J. A. and Matthiessen, J. N. (2004) Developing and refining the biofumigation concept. *Agroindustria* 3: 233–239.

Kruidhof, H. M., Bastiaans, L. and Kropff, M. J. (2009) Cover crop residue management for optimizing weed control. *Plant and Soil* 318: 169-184.

Kuepper, G. (2003) Flea beetle: organic control options. *National Center for Appropriate Technology,* Fayetteville, Arkansas (US). ATTRA publication, 6 pp.

Kuo, S. and Jellum, E. J. (2002) *Influence of winter cover crop and residue management on soil nitrogen availability and corn. Agronomy Journal* 94 (3): 501-508.

Kuo, S. and Sainju, M. (1998) Nitrogen mineralization and availability of mixed leguminous and non-leguminous cover crop residues in soil. *Biology and Fertility of Soils* 26: 346-353.

Langdale, G. W., Blewins, R. L., Karlen, D. L., McColl, D. K., Nearing, M. A., Skidmore, E. L., Thomas, A. W., Tyler, D. D., and Williams, J. R. (1991) Cover crops effects on soil erosion by wind and water. In: Hargrove, W. L. (Ed.), *Cover crops for clean water. Soil and Water Conservation Society of America,* Ankeny (US), 15-22.

Leach, S. S. (1993) Effects of moldboard plowing, chisel plowing and rotation crops on the *Rhizoctonia* disease of white potato. *American Potato Journal* 70: 329-337.

Lehmann, J., Da Silva, J. P. Jr. and Trujillo, L. (2000) Legume cover crops and nutrient cycling in tropical fruit trees production. *Acta Horticulturae* 531: 65-72.

Lewis, J. A. and Papavizas, G. C. (1975) Survival and multiplication of soil-borne plant pathogens as affected by plant tissue amendments. In: Bruehl, G. W. (Ed.), *Biology and Control of Soil-borne Plant Pathogens.* American Phytopathological Society, St. Paul (US), 84-89.

Lewis, J. A. and Papavizas, G. C. (1977) Effect of plant residue on chlamydospore germination of *Fusarium solani* f. sp. *phaseoli* on Fusarium root rot of beans. *Phytopathology* 67: 925–929.

Lewis, W. J., Van Lenteren, J. C., Phatak, S. C., and Tumlinson, J. H. (1997) A total system approach to sustainable pest management. *Proceedings of the Natural Academy of Science* 94: 12243–12248.

Little, S. A., Hocking, P. J. and Greene, R. S. B. (2004) A preliminary study of the role of cover crops in improving soil fertility and yield for potato production. *Communications in Soil Science and Plant Analysis* 35 (3-4): 471-494.

Lopes, C. M., Santos, T. P., Monteiro, A., Rodrigues, M. L., Costa, J. M., and Chaves, M. M. (2011) Combining cover cropping with deficit irrigation in a Mediterranean low vigor vineyard. *Scientia Horticulturae* 129 (4): 603–612.

Lord, J. S., Lazzeri, L., Atkinson, H. J., and Urvin, P. E. (2011) Biofumigation for control of pale potato cust nematodes: activity of Brassica leaf extracts and green manures on *Globodera pallida* in vitro and in soil. *Journal of Agricultural and Food Chemistry* 59: 7882-7890.

Lovett, J. V. and Jessop, R. C. (1982) Effects of residue of crop plant on germination and early growth of wheat. *Australian Journal of Agricultural Research* 33: 909-916.

Lu, Y. C., Bradley Watkins, K., Teasdale, J. R., and Abdul-Baki, A. A. (2000) Cover crops in sustainable food production. *Food Review International,* 16 (2): 121-157.

Lumsden, R. D., Lewis, J. A. and Papavizas, G. C. (1983) Effect of organic amendments on soilborne plant diseases and pathogen antagonists. In: Lockeretz, W. (Ed.), *Environmentally Sound Agriculture.* Praeger Scientific, New York (US), 51-70.

MacDonald, D. H. and Mai, W. F. (1963) Suitability of various cover crops as hosts for the lesion nematode, *Pratylencus penetrans. Phytopathology* 53: 730-731.

Manici, L. M., Lazzeri, L., Baruzzi, G., Leoni, O., Galletti, S., and Palmieri, S. (2000) Suppressive activity of some glucosinolate enzyme degradation products on *Pythium irregulare* and *Rhizoctonia solani* in sterile soil. *Pest Management Science* 56: 921–926.

Manici, L. M., Lazzeri, L. and Palmieri, S. (1997) *In vitro* antifungal activity of glucosinolates and their enzyme derived products towards plant pathogenic fungi. *Journal of Agricultural and Food Chemistry* 45: 2768–2773.

Marks, M. J. and Soane, G. C. (1987) Crop and soil response to subsoil loosening, deep incorporation of phosphorus and potassium fertilizer and subsequent soil management on a range of soil types. Part 1. Response of arable crops. *Soil Use and Management,* 3: 115-122.

Matthiessen, J. N. and Kirkegaard, J. A. (2006) Biofumigation and enhanced biodegradation: opportunity and challenge in soilborne pest and disease management. *Critical Reviews in Plant Sciences,* 25 (3): 235-265.

Mauro, R., Occhipinti, A., Longo, A. M. G., and Mauromicale, G. (2011) Effects of shading on chlorophyll content, chlorophyll fluorescence and photosynthesis of subterranean clover. *Journal of Agronomy and Crop Science,* 197, 57-66.

Mauromicale, G., Occhipinti, A. and Mauro, R. (2010) Selection of shade-adapted subterranean clover species for cover cropping in orchards. *Agronomy for Sustainable Development,* 30: 473-480.

Mauromicale, G., Raccuia, S. A. and Giuffrida, F. (1997) Raccolta e caratterizzazione morfologica e biologica di popolazioni siciliane di trifoglio sotterraneo. *Proceedings of the III Italian National Congress on Biodiversity.* Gallina (RC), June16-17, 89-101.

McGourty, G. (1994) Cover crops for North Coast vineyards. *Practical Winery and Vineyard* 15 (2), 1-6.

McGourty, G. T. and Christensen, L. P. (1998) Cover cropping systems and their management. In: Ingels, C. A., Bugg, R. L., McGourty, G. T., Christensen, L. P. (Eds.), *Cover cropping in vineyards. A grower's handbook.* University of California, Agriculture and Natural Resources Communication Services, Oakland (US), 43-58.

McKenry, M. V. (1991) Marigolds and nematode management. *University of California Kearney Plant Protection Quarterly* 1: 1-4.

McLeod, R. W. and Steele, C. C. (1999) Effects of *Brassica* leaf green manures and crops on activity and reproduction of *Meloidogyne javanica. Nematology* 1: 613–624.

Meisinger, J. J., Hargrove, W. L., Mikkelsen, R. L., Williams, J. R., and Benson, V. W. (1991) Effects of cover crops on groundwater quality. In: Hargrove, W. L. (ed.), *Cover crops for clean water. Soil and Water Conservation Society,* Ankeny, Iowa, 57-68.

Miguez, F. E. and Bollero, G. A. (2005) Review of corn yield response under winter cover cropping systems using meta-analytic methods. *Crop Science* 45 (6): 2318-2329.

Mitchell, W. H. and Tell, M. R. (1997) Winter annual cover crops for no-tillage corn production. *Agronomy Journal* 69, 569-573.

Morgan, R. P. C. and Duzant, J. H. (2008) Modified MMF (Morgan–Morgan–Finney) model for evaluating effects of crops and vegetation cover on soil erosion. *Earth Surface Processes and Landforms* 33 (1): 90–106.

Moser, F. (1942) Influence of leguminous plant additions on the organic matter content and available nutrient supply of southern soil. *Journal of American Society of Agronomy* 34: 711-719.

Muller, J. (1999) The economic importance of *Heterodera schachtii* in Europe. *Helminthologia* 36: 205–213.

Muller, M. M., Sundman, V., Soinvaara, O., and Merilainen, A. (1988) Effect of chemical composition on the release of nitrogen from agricultural plant materials decomposing in soil under field conditions. *Biology and Fertility of Soils* 6: 78-83.

Nair, A. and Ngouajio, M. (2012) Soil microbial biomass, functional microbial diversity, and nematode community structure as affected by cover crops and compost in an organic vegetable production system. *Applied Soil Ecology,* 58: 45-55.

Nicholls, C. I. and Altieri, M. A. (2012) Plant biodiversity enhances bees and other insect pollinators in agroecosystems. A review. *Agronomy for Sustainable Development, in press* [DOI 10.1007/s13593-012-0092-y].

Occhipinti, A., Mauro, R., Ierna, A., Restuccia, A., and Mauromicale, G. (2007) Sul possibile contributo di due leguminose autoriseminanti come "cover crop" nel vigneto. *Proceedings of the XXXVII SIA (Società Italiana di Agronomia) annual meeting,* Catania, 13-14 settembre, 295-296.

Oleszek, W. and Jurzysta, M. (1987) The allelopathic potential of alfalfa root medicagenic acid glycosides and their fate in soil environments. *Plant and Soil* 98: 67-80.

Pagliai, M., Raglione, M., Panini, T., Maletta, M., and La Marca, M. (1995) The soil structure after ten years of conventional and minimum tillage of two Italian soils. *Soil and Tillage Research,* 34: 209-223.

Pecetti, L. and Piano, E. (2002) Variation of morphological and adaptative traits in subterranean clover populations from Sardinia (Italy). *Genetic Resources and Crop Evolution* 49: 189-197.

Phatak, S. C., 1998. Managing pests with cover crops. In: US Sustainable Agricultural Network (Ed.), *Managing cover crop profitability.* University of Vermont, Burlington (US), 25-33.

Pieters, A. J. (1927) *Green manuring - principles and practice.* New York: John Wiley, 356 pp.

Porter, R. (1988) Establishing vineyard cover crops. The Australian Grapegrower, *Winemaker,* 422: 18-19.

Power, J. F. (1990) Legumes and crop rotations. In: Francis, C. A. (Ed.), *Sustainable agriculture in temperate zone.* John Wiley and Sons, New York (US), 178-204.

Putnam, A. R. (1990) Vegetable weed control with minimal herbicide inputs. *HortScience* 25: 155-158.

Putnam, A. R. and De Frank, J. (1982) Use of phytotoxic plant residues for selective weed control. *Crop Protection* 2: 173-181.

Quemada, M. and Cabrere, M. L. (1995) Carbon and nitrogen mineralized from leaves and stems of four cover crops. *Soil Science Society of America Journal* 59: 471-477.

Ranells, N. N. and Wagger, M. G. (1997) Nitrogen-15 recovery and release by rye and crimson clover cover crops. *Soil Science Society of America Journal* 61: 943-948.

Rask, L., Andreasson, E., Ekbom, B., Eriksson, S., Pontoppidan, B., and Meijer, J. (2000) Myrosinase: gene family evolution and herbivore defense in Brassicaceae. *Plant Molecular Biology* 42: 93–113.

Reddy, K. C., Soffes, A. R., Prine, G. M., and Dunn, R. A. (1986) Tropical legumes for green manure: Nematode populations and their effects on succeeding crop yield. *Agronomy Journal* 78: 5-10.

Rod, P. (1977) Observations sur le ruissellements dans le vignoble vadnais. *Bulletin de l'Association Romande pour la Protection des Eaux et de l'Air,* 82: 41-44.

Rodríguez-Kábana, R., Robertson, D. G., Wells, L., and Young, R. W. (1988) Hairy indigo for the management of *Meloydogine arenaria* in peanut. *Nematropica* 18: 137-142.

Rosa, E. A. S. and Rodrigues, P. M. F. (1999) Towards a more sustainable agricultural system: The effect of glucosinolates on the control of soilborne diseases. *Journal of Horticultural Science and Biotechnology* 74: 667–674.

Salmeron, M., Isla, R. and Cavero, J. (2011) Effect of winter cover crop species and planting methods on maize yield and N availability under irrigated Mediterranean conditions. *Field Crops Research* 123 (2): 89-99.

Sarrantonio, M. (2000) Building soil fertility and tilt with cover crops. In: US Sustainable Agricultural Network (Ed.), *Managing cover crop profitability,* University of Vermont, Burlington (US), 16-24.

Schenk, S. U. and Werner, D. (1991) Beta-(3-isoxazolin-5-on-2yl)-alanine from *Pisum:* allelopathic properties and antimycotic bioassay. *Phytochemistry* 30: 467 – 470.

Shanks, L. W., Moore, D. E. and Epifanio Sanders, C. (1998) Soil erosion. In: Ingels, C. A., Bugg, R. L., McGourty, G. T., Christensen, L. P. (Eds.), *Cover cropping in vineyards. A grower's handbook.* University of California, Agriculture and Natural Resources Communication Services, Oakland (US), 80-85.

Shennan, C. (1992) Cover crops, nitrogen cycling and soil properties in semi-irrigated vegetable production systems. *HortScience* 27: 749-754.

Shilling, D. G., Liebl, A. D. and Worsham, R. A. (1985) Rye (*Secale cereale* L.) and wheat (*Triticum aestivum* L.) mulch: the suppression of certain broadleaved weeds and the isolation and the identification of phytotoxins.

In: Thompson, A. C. (Ed.), *The chemistry of allelopathy: Biochemical interactions among plant.* Amer. Chem. Soc., Washington D.C. (US).

Smith, H. and Morgan, D. C. (1983) The function of phytochrome. In: *Phytomorphogenesis. Encyclopedia of Plant physiology,* New Series. Shropshire, W. e Mohr, H. (Eds.), Springer- Verlag, New York (US), 491-517.

Snodgrass, G. L., Scott, W. P. and Smith, J. W. (1984) Host plants and seasonal distribution of the tarnished plant bug (Hemiptera: Miridae) in the Delta of Arkansas, Louisiana, and Mississippi. *Environmental Entomology* 13: 110 –116.

Stadelbacher, E. A. (1981) Role of early-season wild and naturalized host plants in the build-up of the F_1 generation of *Heliothis zea* and *H. virescens* in the Delta of Mississippi. *Environmental Entomology* 10: 766 –770.

Sullivan, P. (2003) Overview of cover crops and green manures. *National Center for Appropriate Technology,* Fayetteville, Arkansas (US). ATTRA publication, 16 pp.

Talamucci, P. (2002) Trifoglio sotterraneo (*Trifolium subterraneum* L.; *T. brachycalycinum* Katzn. e Morley; T. *yanninicum* Katzn. e Morley). In: Baldoni, R., Giardini, L. (Coord.), *Coltivazioni erbacee - Foraggere e tappeti erbosi,* vol. 3, pp. 153-160. Pàtron Ed., Bologna (Italy).

Teasdale, J. R. (1996) Contribution of cover crops to weed management in sustainable agricultural systems. *Journal of Production Agriculture* 9: 475-479.

Teasdale, J. R. and Abdul-Baki, A. A. (1995) Soil temperature and tomato growth associated with black polyethylene and hairy vetch mulches. *Journal of the American Society of Horticultural Science,* 120: 848-853.

Teasdale, J. R. and Mohler, C. L. (1993) Light transmittance, soil temperature and soil moisture under residue of hairy vetch and rye. *Agronomy Journal,* 85: 673-680.

Theophrastus. Enquiry into plants. *Loeb Classical Library,* 1916.

Thorup-Kristensen, K., Magid, J. and Jensen, L. S. (2003) Catch crops and green manures as biological tools in nitrogen management in temperate zones. *Advances in Agronomy* 51: 227–302.

Trdan, S., Valič, N., Žnidarčič, D., Didrih, M., Bergant, K., Zlatič, E., and Milevoj, L. (2005) The role of Chinese cabbage as a trap crop for flea beetles (Coleoptera: Chrysomelidae) in production of white cabbage. *Scientia Horticulturae* 106: 12-24.

Trott, D., Lepage, J. and Hebert, V. R. (2012) Assessing natural isothiocyanate air emissions after field incorporation of mustard cover crop. *Bulletin of Environmental Contamination and Toxicology* 88 (3): 482-485.

Van den Boogert, P. H. J., Velvis, H., Ettema, C. H., and Bouwman, L. A. (1994) The role of organic matter in the population dynamics of the endoparasitic nematophagous fungus *Drechmeria coniospora* in microcosms. *Nematologica* 40: 249-257.

Vigil, M. F. and Kissel, D. E. (1991) Equations for estimating the amount of nitrogen mineralized from crop residue. *Soil Science Society of America Journal* 55: 757-761.

Wagger, M. G. (1989) Cover crop management and nitrogen rate in relation to growth and yield of no-till corn. *Agronomy Journal* 81: 533-538.

Wagner-Riddle, C., Gillespie, T. J. and Swanton, C. J. (1994) Rye cover crop management impact on soil water content, soil water content, soil temperature and soybean growth. *Canadian Journal of Plant Science,* 74: 485-495.

Wang, K. H., Sipes, B. S. and Schmitt, D. P. (2002) Management of *Rotylenchulus reniformis* in pineapple, *Ananas comosus*, by intercycle cover crops. *Journal of Nematology* 34: 106-114.

Warner, G. (1999) Cover crop advantages. *Goodfruit Grower* 50 (9): 10.

Westerdahl, D. A., Caswell-Chen, E. P. and Bugg, R. L. (1998) Nematodes. In: Ingels, C. A., Bugg, R. L., McGourty, G. T., Christensen, L. P. (Eds.), *Cover cropping in vineyards. A grower's handbook.* University of California, Agriculture and Natural Resources Communication Services, Oakland (US), 113-125.

Weston, L. A., Harmon, R. and Mueller, S. (1989) Allelopathic potential of sorghum sudangrass hybrid (Sudex). *Journal of Chemical Ecology* 15: 1855 – 1865.

White, R. H., Worsham, A. D. and Blum, U. (1989) Allelopathic potential of legume debris and aqueous extracts. *Weed Science* 37: 674-679.

Wiedenfeld, B., Fenn, L. B., Miyamoto, S., Swietlik, D., and Marlene C. (1999) Using sod to manage nitrogen in orchard floors. *Communications in Soil Science and Plant Analysis* 30: 353 – 363.

Williams, W. A., Graves, W. L. and Cassman, K. G. (1990) Nitrogen fixation by irrigate berseem clover versus soil nitrogen supply. *Journal of Agronomy and Crop Science* 164: 202-207.

Wolpert, J. A., Phillips, P. A., Striegler, R. K., McKenry, M. V., and Foott, J. H. (1993) Berber orchardgrass tested or cover crop in commercial vineyard. *California Agriculture* 47: 23-25.

Workneh, F., Van Bruggen, A. H. C., Drinkwater, L. E., and Shennan, C. (1993) Variables associated with corky root and *Phytophthora* root rot of tomatoes in organic and conventional farms. *Phytopathology* 83: 581–589.

Wyland, L. J., Jackson, L. E. and Schulbach, K. F. (1995) Soil plant nitrogen dynamics following incorporation of a mature rye cover crop in a lettuce production system. *Journal of Agricultural Science* 124: 17-25.

Yeates, G. W. (1987) How plants affect nematodes. *Advances in Ecological Research* 17: 61-113.

Zasada, I., Rice, C. P. and Meyer, S. L. F. (2007) Improving the use of rye (*Secale cereale*) for nematode management:potential to select cultivars based on *Meloidogyne incognita* host status and benzoxazinoid content. *Nematology* 9: 53-60.

Zohary, D. and Katznelson, J. (1958) Two species of subterranean clover in Israel. *Australian Journal of Botany* 6 (2): 177-182.

In: Weeds and Their Ecological Functions
Editor: Alireza Taab
ISBN: 978-1-62808-332-3

Chapter 6

Beneficial and Detrimental Interactions Between Weeds and Other Pests of Sugarcane

*Allan T. Showler**
USDA-ARS, Kerrville, Texas, US

Abstract

Although weeds are commonly regarded as being negative influences on crop production, they also contribute toward vegetational diversity in agroecosystems which generally promotes food web stability. Di- and polyculture crop systems have been shown to reduce populations and injury inflected by many crop pests, and weeds have been shown to induce similar effects. Sugarcane, being a perennial plant, offers an illustrative example for the effects of weed communities in and around the field, and sugarcane is arguably one of the most widely studied crops in terms of its relationships with weeds. This review examines what is known about weed interactions with other biotic compartments of sugarcane agroecosystems, including nematodes, plant pathogens, arthropod communities, and the crop itself. Three economically important

* Correspondence: Allan T. Showler, Ph.D., USDA-ARS, 2700 Fredericksburg Road, Kerrville, Texas 78028 USA. Tel: (830) 792-0319, E-mail: allan.showler@ ars.usda.gov; Mention of trade names or commercial products in this publication is solely for the purpose of providing specific information and does not imply recommendations or endorsement by the U.S. Department of Agriculture.

stalkboring lepidopteran pests of sugarcane, in particular, have received the most attention: the sugarcane borer (*Diatraea saccharalis* [F.]), and the Mexican rice borer, *Eoreuma loftini* (Dyar), the eldana borer (*Eldana saccharina* Walker). *Diatraea saccharalis* interactions with weeds involve enhanced populations of predators, especially the red imported fire ant, *Solenopsis invicta* Buren. *Eoreuma loftini* and *E. saccharina* associations with weeds and other crops occur more on an agricultural landscape scale, wherein plant diversification can exacerbate the spread and degree of infestation. In terms of sugarcane protection, weeds are detrimental as competitors with the crop and as reservoirs for diseases, arthropod pests (particularly during winter), and vectors of sugarcane diseases. In instances where alternate plant hosts of sugarcane pests are problematic, area-wide strategies that account for the effects of alternate host plants on pest populations should be developed to minimize crop injury. It will be necessary to consider the interactions of weeds with sugarcane and its pests on a regional basis in order to address interactions peculiar to each region.

1. Introduction

In agriculture, the term "weeds" refers to plants that do not constitute the crop and are unwanted or undesirable as competitors (Harper, 1977), reservoirs for plant pathogens and arthropod pests (Allen and Smith, 1958; Brandenburg and Kennedy, 1981), and as obstructive to harvest operations (Fisher et al., 1977; McWhorter and Anderson, 1981). Weeds, however, also contribute toward vegetational diversity, conferring structural and biochemical complexity to agroecosystems and area-wide agricultural landscapes. Vegetational complexity is often associated with faunal diversity (Smith and Whittaker, 1980; Barney et al., 1984; Showler and Reagan, 1991), and overall species diversity is frequently linked with ecosystem stability (Graham, 1956; Harper, 1969; Odum, 1971; Way, 1977) and reduced incidence of pest outbreaks (Pimentel, 1961; Gibson and Jones, 1977).

By supporting diversified arthropod populations, vegetational diversification has been reported to suppress arthropod pest populations and their injury to crops. Cucumber plants (*Cucumis* sativa L.) for example, had 75% fewer striped cucumber beetles (*Acalymma vittata* [F.]) in cucumber/corn (*Zea mays* L.)/broccoli (*Brassica oleracea* L. var. *botrytis* L.) polycultures than in cucumber monocultures (Bach, 1980). Similarly, corn/bean (*Phaseolus* sp.) dicultures were less injured by the banded cucumber beetle (*Diabrotica balteata* LeConte), fall armyworm (*Spodoptera frugiperda* Smith), Mexican bean beetle (*Epilachna varivestis* Mulsant), and the green leafhopper (*Empoasca kraemeri* [Ross and Moore]) than corn monocultures (Altieri et al., 1978, Coll and Bottrell, 1995). Cover crops, such as beans in apple (*Malus domestica* Borkh.) orchards, have been associated with reduced codling moth (*Cydia pomonella* L.), rosy apple aphid (*Anuraphis roseus* Baker), and cicadellid (leafhopper) infestations (Altieri and Schmidt, 1986). Modified applications of polyculture include "strip cropping" for building natural enemy populations by planting alfalfa (*Medicago sativa* L.) in corn fields for example, and crop rotations that capitalize on vegetational diversification on a temporal basis (Laster, 1974; Luckman and Metcalf, 1975; Reynolds et al., 1975).

Weeds have also been associated with suppression of insect pests. Examples abound, but include protection from green stinkug (*Nezara viridula* L.), fall armyworm, and corn earworm (*Heliothis zea* Boddie) in soybeans (*Glycine max* [L.] Merr.), potato leafhopper (*Empoasca fabae* [Harris]) in alfalfa; and cabbage butterfly (*Pieris rapae* L.), cabbage aphid (*Brevicoryne brassicae* L.), and cabbage whitefly (*Aleyrodes brassicae* [Walker]) in brussels sprouts (*Brassica oleracea* L. var. *gemmifera* DC) (Dempster, 1969; Smith, 1969; Altieri and Whitcomb, 1979; Ryan et al., 1980; Altieri, 1981; Altieri and Todd, 1981; Altieri, 1984; Oloumi-Sadeghi et al., 1987).

Several theories have been developed to explain the reduction of pest infestations and crop injury in vegetationally diversified crop systems. One involves spatial dilution of the primary resource (*e.g.*, the crop, rendering it less available) (Altieri et al., 1977; Risch, 1979; Letourneau, 1986). Another involves chemical or structural interference with host finding and utilization by pests (Atsatt and O'Dowd, 1976; Risch, 1979, 1980), and a third theory is that diversification contributes toward enhanced populations of natural enemies (Altieri and Whitcomb, 1980; Barney et al., 1984; Foster and Ruesink, 1984; House, 1989; Taylor et al., 2006). Sugarcane is a crop in which weed interactions with other biotic agroecosystem components have been explored in greater depth than many other field crop systems. Hence, sugarcane was chosen for this review in order to show how weeds can exert both deleterious and beneficial effects on crop production and protection.

2. SUGARCANE

Sugarcane (*Saccharum* spp.) is a perennial crop cultivated on roughly 20 million hectares in tropical and subtropical regions of more than 100 countries

(FAO, 2009; March, 2006; GMO Compass, 2008). Some 1,325 million tons of sugarcane are harvested each year for production of sugar, rum, pharmaceuticals and other chemicals, and energy (*i.e.*, biofuel) (Sharpe, 1998; Marsh, 2006; Nass et al., 2007; Outlaw et al., 2007; Salassi, 2008). The plants are vegetatively propagated from cuttings and whole stalks, and after each harvest the regrowth, called ratoons, grows from the stubble. Like most crops, sugarcane is vulnerable to an array of abiotic and biotic factors (*e.g.*, weed competition, pathogen infection, arthropod injury, low winter temperatures, poor soil aeration and drainage, and weak or vulnerable genotypes) that detract from optimal production potential (Edgerton, 1939; Rands and Dopp, 1938; Carter, 1977; Hoy and Schneider, 1988b). Although as many as 20 ratoon crops can be obtained from a single planting (Blackburn, 1984), those factors are also associated with deterioration of productivity from one season to the next, termed "ratoon decline," sometimes limiting production to only two ratoon seasons (Edgerton, 1939; Hoy and Schneider, 1988a,b).

3. Weeds Compete with the Crop

Infestation by weeds, particularly the perennial johnsongrass (*Sorghum halepense* [L.] Pers.), in sugarcane systems can cause substantial yield losses (McWhorter, 1972; Ali et al., 1986; Ogunwolu et al., 1987) of up to 84% (Arevelo et al., 1977). Sugarcane biomass and sugar yields from fields heavily infested with johnsongrass were reduced by 36% and 31%, respectively, compared with weed-free systems (Ali et al., 1986). Weeds compete more effectively with sugarcane than many short-season row crops partly because of the wide sugarcane row spacing (≈1.8 m) and relatively slow seedling growth (Peng, 1984). In large mechanized sugarcane systems such as those of Florida and Louisiana, weeds are controlled using cultivation and herbicides (Ogunwolu et al., 1987), but most weed species are not highly competitive, frequently being annuals that die back after canopy closure (Ali and Reagan, 1985a). Showler and Reagan (1991), however, reported that typical stands of annual summer weeds were associated with reductions of as much as 24% in sugarcane stalk density, 19% of biomass, and 15% of commercial sugar production. Purple nutsedge (*Cyperus rotundus* L.) at low (58-246 shoots/m^2), medium (318-773 shoots/m^2), and high (675-1,198 shoots/m^2) densities caused sugarcane yield reductions of 13.5%, 29.3%, and 45.2%, respectively (Durigan, 2005). Bermudagrass (*Cynodon dactylon* L. Pers.) diminished plant-season sugarcane stalk densities by 13–23%, the first ratoon by 8–15%, and the second ratoon by 8–10%, and sugar yield reductions were 8–32% in the plant cane and 9% in each of the following two ratoon years (Richard and Dalley, 2007).

Weeds in sugarcane fields of relatively wet regions, such as southern Louisiana, do not appear to compete with the crop for water, a conclusion supported by low free proline concentrations (high accumulations of which indicate water deficit stress) in sugarcane stalks (Showler et al., 1990; Showler, 2002, 2012; Showler and Castro 2010a). After canopy closure, weeds die and form a matting, or mulch, of dry vegetation on the soil surface that conserves soil moisture, also indicated by low levels of free proline (Showler et al., 1990; Showler, 2002). Effects of weeds on soil fertility of agricultural systems are well-documented (Hatfield and Stewart, 1997; Zimdahl, 2004), but in the context of sugarcane production, examples are limited to a study in India showing that sugarcane weeds removed 64 kg of nitrogen per ha (Yadav, 1985).

4. WEEDS AND PATHOGENS OF SUGARCANE

Sugarcane mosaic virus can infect all sugarcane cultivars, and of the 69 aphid species found on sugarcane of Louisiana, sugarcane mosaic virus can be vectored by the corn leaf aphid (*Rhopalosiphum maidis* [Fitch]), pea aphid (*Acythosiphon pisum* [Harris]), red sowthistle aphid (*Dactynotus abrosiae* [Thomas]); greenbug (*Schizaphus graminum* [Rondani]), rusty plum aphid (*Hysteroneura setariae* [Thomas]), and sedge aphid (*Carolinaia cyperi* Ainslie) (Brandes, 1920, Ingram and Summers, 1936; Tate and Vanderberg, 1939; Abbott and Charpentier, 1963; Komblas, 1964; Abbott and Tippett, 1966; Benda, 1969).

Grassy weeds that serve as alternate hosts of the corn leaf aphid include johnsongrass, maidencane (*Panicum hemitomon* Schult), jungle rice (*Echinochloa colonum* [L.] Link), and crabgrass (*Digitaria sanguinalis* [L.] Scop.), the pea aphid on white sweetclover (*Melilotus alba* Dest.), and many other weed species; the green sowthistle aphid and the red sowthistle aphid on spiny sowthistle (*Sonchus asper* [L.] All.), the greenbug on *Digitaria* spp. and many other weed species; and the sedge aphid on *Cyperus* spp. (Ingram and Summers, 1938; Ingram et al., 1951; Komblas, 1964).

Further, sugarcane mosaic virus can infect an array of graminaceous crops and weeds, including pearl millet (*Pennisetum glaucum* [L.] R. Br.), maiden grass (*Miscanthus sinensis* Anderss.), bull crowngrass (*Paspalum boscianum*

Flugge), crabgrass; yellow foxtail (*Setaria lutescens* [Weigel] Hubb.), giant bristlegrass (*S. magna* Griseb.), barnyardgrass (*Echinochloa crus-galli* [L.] Beauv.), fall panicgrass (*Panicum dichtomiflorum* Michx.), broadleaf signalgrass (*Brachiaria platyphylla* Nash.), johnsongrass, itchgrass (*Rottboellia exaltana* L.), and Saint Augustine grass (*Stenotaphrum secundatum* [Walt.] Kuntze) (Edgerton, 1959; Todd, 1964; Forbes et al., 1967; Gillespi and Koike, 1973). A study comparing weedy sugarcane plots against plots where weeds were eliminated revealed that sugarcane mosaic virus infection of sugarcane plants was as much as 14% greater in the weedy plots (Showler, 1987).

Sugarcane yellow leaf virus is transmitted by the yellow sugarcane aphid (*Melanaphis sacchari* [Zehntner]) (Akbar et al., 2010; Paray et al., 2011). The yellow sugarcane aphid is hosted by crop and weed hosts in the genera *Echinochloa*, *Oryza*, *Panicum*, *Pennisetum*, *Saccharum*, and *Sorghum* (Denmark, 1988), and sugarcane yellow leaf virus is hosted by plants of the graminaceous Andropogonineae subtribe (Scaglusi and Lockhart, 2000), comprised of >75 genera.

The bacterium *Leifsonia xyli xyli*, known as ratoon stunting disease, is found in graminaceous weeds such as para grass (*Brachiaria mutica* Stapf), Rhodes grass (*Chloris gayana* Kunth), bermudagrass, jungle rice, Guinea grass (*Panicum maximum* Jacq.), Natal grass (*Rhynchelytrum repens* [Wiled.] Hubb.), and rat-tail grass (*Sporobolus capensis* Chiov.) (Steindl, 1957). Although ratoon stunting disease persists between growing seasons in the soil, the roles of weeds in the pathogen's transmission to sugarcane and as between-season hosts, have not been reported.

Another soil-borne pathogen of sugarcane, *Pythium arrhenomanes*, also infects weeds including bermudagrass; broadleaf signalgrass; browntop panicum (*Panicum ramosum* L.), barnyardgrass, crabgrass, itchgrass, goosegrass (*Eleusine endica* [L.] Gaertn.), johnsongrass, Italian ryegrass (*Lolium multiflorum* Lam.), and purple nutsedge (*Cyperus rotundus* L.) (Dissayanake et al., 1997).

5. Weeds and Phytophagous Nematodes of Sugarcane

Sugarcane supports at least 14 phytophagous nematode genera, some of which are major pests, including root knot nematode, (*Meloidogyne* spp.), lesion nematode (*Pratylenchus* spp.), stubby root nematode (*Trichodorus* spp.), and stunt nematode (*Tylenchorhynchus* spp.) (Birchfield, 1984). Each of the phytophagous nematode genera can inhibit host plant water and nutrient uptake (Astudillo and Birchfield, 1980; Ramirez, 1981; Wilcox-Lee and Loria, 1987). Nematode control tactics, where practiced, involve fallow-plowing, crop rotation, and nematicides (Birchfield, 1972), but application of some nematicides, such as aldicarb, can reduce natural enemies of the sugarcane borer (*Diatraea saccharalis* [F.], resulting in heightened stalkborer damage to the crop (Showler and Reagan, 1991; Showler et al., 1991). Many weed species found in sugarcane fields host agriculturally important nematodes (Gaskin, 1958; Golden and Birchfield, 1965; Hogger and Bird, 1974; Bendixon et al., 1979; Manuel et al., 1980). Although weed management might contribute toward control of nematodes (Peng, 1984), the only nematode detected in greater numbers in weed-infested systems than in bare soil systems was *Criconemella* spp. (Showler et al., 1990) late in the first growing season, but *Meloidogyne* spp. and *Tylenchorhynchus annulatis* (Cassidy) Golden populations were smaller (by as much as 79% for *T. annulatis*) in the sugarcane rhizosphere in weedy field plots than in plots where weeds had been eliminated (Showler et al., 1990).

6. Weeds and Three Selected Stalkborers

Many species of stalk boring insects attack sugarcane, but *D. saccharalis, E. loftini*, and *E. saccharina* were selected as examples for this chapter because each of them have been shown to interrelate with weeds (Showler and Reagan, 2012). Differences in how each of the three sugarcane stalkboring pests interact with weeds demonstrate a variety of influences that vegetational diversity can exert on pests of the same guild, and how those influences affect crop protection.

6.1. Sugarcane Borer (*Diatraea Saccharalis* [F.], Lepidoptera: Crambidae)

Diatraea saccharalis originated in parts of the Western Hemisphere's tropics, and the pest invaded the United States after arrival in Louisiana in about 1855 (Capinera, 2001; Showler and Reagan, 2012). The insect currently inhabits a range encompassing the United States' Gulf Coast, the Caribbean,

Central America, and South America to subtropical regions of northern Argentina (Capinera, 2001).

While it is a major pest of corn, sorghum (*Sorghum bicolor* [L.] Moench), and sugarcane in most of Latin America, *D. saccharalis* is the chief pest of corn in Argentina (Moré et al., 2003). In northern Tamaulipas (northeastern Mexico), Texas, and Louisiana, larvae that survive winter pupate and resulting adults become active by spring, followed by four or five nonsynchronous generations through the fall (Reagan and Martin, 1989; Rodriguez-del-Bosque et al., 1995).

Completion of a summertime generation requires ≈25–40 days whereas >200 days are needed during winter (Reagan and Martin, 1989; Capinera, 2001). Populations infesting corn in northern Tamaulipas decline through winter, and substantial mortality occurs when temperatures of <-5° C last for several hours (Rodriguez-del-Bosque et al., 1995).

Eggs are deposited on exposed upper and lower surfaces of green leaf blades near the top of the plant, making them vulnerable to egg parasites (Browning and Melton, 1987). Hatched larvae tend to feed on leaf tissue or tunnel inside the midrib, and second and third instars bore into the stalk (Reagan and Martin, 1989).

After 25–30 days in warm weather, and 30–35 days during cool weather, larvae reach the pupal stage (Capinera, 2001). During cold winters, larval development stops in response to the low temperature (Capinera, 2001) and possibly because of temperature- and photoperiod-related diapause (Fuchs et al., 1979). Pupae are protected from external biological and chemical mortality factors by a thin layer, or "window," of plant tissue. In 8–9 days (up to 22 days under cool conditions), adults emerge from the pupae and break through the window.

The inner whorl of younger plants can be killed causing a condition called "dead heart" and the tops of mature infested plants tend to weaken, die, or lodge (Reagan and Martin, 1989). Infestations diminish the quantity and purity of sugarcane juice, resulting in sucrose yield declines of 10–20% (Capinera, 2001). *Diatraea saccharalis* (Charpentier et al., 1967; Pollet et al., 1978, White et al., 2004) is responsible for >90% of all insect-related sugarcane losses in Louisiana (Hensley, 1971; Ogunwolu et al., 1987) (in the Lower Rio Grande Valley of Texas, however, the intrusion of *E. loftini* beginning in the early 1980s supplanted the sugarcane borer as the key pest of sugarcane [Legaspi et al., 1997]).

Diatraea saccharalis tunneling also provides portals of entry by red rot (*Colletrotrichum falcatum* Went) and other sugarcane stalk rot fungi

(Ogunwolu et al., 1991) into growing plants and into tunneled sugarcane seed pieces (Capinera, 2001). Ulloa et al. (1982) indicated that sugarcane yield losses increase as infestation intensity increases, and calculated that an average of one bored internode per stalk causes the loss of 2.5 kg of sugar per ton of harvested cane.

Diatraea saccharalis is an occasional pest of other graminaceous crops, such as corn (although in Argentina it is the key pest of corn [Moré et al., 2003]), rice (*Oryza sativa* L.); sorghum, and sudangrass (*Sorghum bicolor* [L.] Moench spp. *drummondi* [Nees ex Steud.] de Wet and Harlan) (Fuller and Reagan, 1989; Sosa, 1990; Moré et al., 2003; Lv et al., 2008), used as livestock forage. Larvae are not commonly encountered in winter corn, but sugarcane and large-diameter stems of some grassy weeds, such as johnsongrass, *Paspalum* sp., *Panicum* spp., *Holcus* sp., and *Andropogon* sp., are important overwintering sites (Capinera, 2001). Stalkborers are known to move from weeds hosts to sugarcane (Nagarkatti and Nair, 2001), and alternate hosts of *D. saccharalis* are common to sugarcane growing areas, including johnsongrass (Table 1).

Diatraea saccharalis inflicts up to 69% more damage to sugarcane in southern Louisiana where non-selective, persistent pesticides, such as aldicarb, chlordane, heptachlor, and mirex have suppressed natural enemy populations than nontreated sugarcane (Long et al., 1958; Hensley et al., 1961; Charpentier et al., 1967; Negm and Hensley, 1967, 1969; Reagan et al., 1972; Showler and Reagan, 1991).

Louisiana sugarcane fields infested with monocot, dicot, or monocot-dicot mixtures, however, were more diverse (Shannon diversity index) in terms of soil surface- and foliage-associated arthropods, including natural enemies of *D. saccharalis*, than bare-soil systems (Ali and Reagan, 1985a). Sugarcane in Louisiana supports ≥84 spider species in 18 families (Ali and Reagan, 1985b) (Table 3), and spiders are important egg predators of *D. saccharalis*, second only to the imported fire ant (*Solenopsis invicta* Buren) (Negm and Hensley, 1969). Weed growth (grasses and broadleaves) has been associated with heightened spider diversity (Hatley and McMahon, 1980; Reichert and Lockley, 1984; Ali and Reagan, 1986) and abundances on the soil surface and on sugarcane foliage (Showler and Reagan, 1991).

Predators can be a decisive factor to the control of *D. saccharalis*, particularly in regions where *S. invicta* thrives, such as in Louisiana (Showler and Reagan, 1991).

Table 1. Crop and non-crop host plants of *Diatraea saccharalis*, *Eoreuma loftini*, and *Eldana saccharina* (Showler et al., 2011; Showler and Reagan, 2012)

Stalkborer species	Host plant	
	Common name	Scientific name
Diatraea saccharalis	Corn	*Zea mays* L.
	Rice	*Oryza sativa* L.
	Sorghum	*Sorghum bicolor* (L.) Moench
	Sugarcane	*Saccharum* spp. and hybrids
		Andropogon spp.
	Johnsongrass	*Sorghum halepense* (L.) Pers.
		Holcus spp.
		Panicum spp.
		Paspalum spp.
	Sudangrass	*Sorghum bicolor* (L.) Moench ssp.
		drummondi (Nees ex Steud.) de Wet and
		Harlan
	Barley	*Hordeum vulgare* L.
	Corn	*Zea mays* L.
	Rice	*Oryza sativa* L.
	Sorghum	*Sorghum bicolor* (L.) Moench
	Sugarcane	*Saccharum* spp. and hybrids
	Wheat	*Triticum aestivum* L.
	Amazon sprangletop	*Leptochloa panicoides* P. Beauv.
	Barnyardgrass	*Echinochloa crus-galli* (L.)
	Bermudagrass	*Cynodon dactylon* (L.) Pers.
	Bristlegrass	*Setaria lutescens* (Weigel) Hubb.
	Broadleaf signalgrass	*Urochloa platyphylla* (Munro ex C. Wright)
	Broom corn-millet	*Panicum miliaceum* L.
	Bulrush	*Scripus* sp.
	Canna	*Canna* sp.
	Johnsongrass	*Sorghum halepense* (L.) Pers.
	Lemongrass	*Cymbopogon citratus* (DC, ex Nees) Stapf
	Pampasgrass	*Cortaderia selloana* (Schult. and Schult. F.)
	Panicum grass	*Panicum* sp.
	River reed	*Arundo donax* L.
	Sudangrass	*Sorghum bicolor* (L.) Moench ssp.
		drummondi (Nees ex Steud.) de Wet and
		Harlan
	Vasey's grass	*Paspalum urvillei* Steud.
Eoreuma	__	*Bromus* sp.
loftini	__	*Lolium* sp.

Stalkborer species	Host plant	
	Common name	Scientific name
Eldana saccharina	Cassava	*Manihot esculenta Crantz*
	Corn	*Zea mays L.*
	Millet	*Eleusine coracana (L.) Gaertn.*
	Rice	*Oryza sativa L.*
	Sorghum	*Sorghum bicolor (L.) Moench*
	Sugarcane	*Saccharum spp. and hybrids*
	Bulrush	*Typha latifolia L.,*
	Pigweed	*Amaranthus dubius (Mart. ex Thell.)*
	Sedges	*Cyperus papyrus L., C. fastigiatus Rottb., and*
		C. dives Delile
	Jaraguagrass	*Hyparrhenia rufa (Nees) Stapf,*
	Guineagrass	*Urochloa maxima (Jacq.) R. Webster*
	Elephant grass	*Pennisetum purpureum Schumach.*
	Itchgrass	*Rottboellia exaltana L.*
	Wild sorghum	*Sorghum bicolor (L.) Moench ssp. arundinaceum* de Wet and Harlan

Table 2. Ant species collected from sugarcane fields in southern parishes of Louisiana, and Cameron, Hidalgo, and Willacy counties, Texas (Showler et al., 1990; Showler and Reagan, 1991; Breene et al., 1993)

Cardiocondyla minutoir Forel
Crematogaster clara Mayr
Foreslius sp.
Hypoponera opaciceps (Mayr)
Iridomyrmex humilis Mayr
Iridomyrmex pruinosus (Roger)
Monomorium minimum (Buckley)
Monomorium pharaonis (L.)
Paratrechina melanderi Wheeler
Paratrechina vividula (Nylander)
Pheidole sp. flavus group
Pheidole dentata Mayr
Pheidole moerens Wheeler
Pogonomyrmex barbatus (F. Smith)
Ponera opaciceps Mayr
Ponera trigona opacior Forel
Solenopsis geminata (F.)
Solenopsis invicta Buren
Solenopsis saevissima richteri Forel
Solenopsis xyloni McCook
Strumigenys louisianae Roger
Tetramorium guineense (L.)

Ants were collected using pitfall traps, sweep nets, aspirators, and by hand.

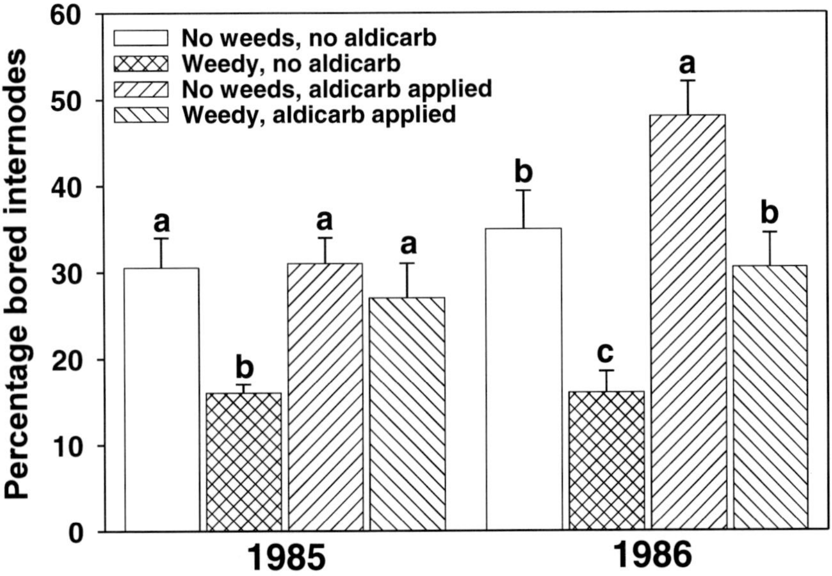

Figure 1. Mean (± SE) percentage bored sugarcane stalk internodes in field plots where weeds were eliminated using pre-plant herbicide and manual weed removal occurred but early-season soil-incorporated aldicarb was not applied, weeds were permitted to grow but aldicarb was not applied, weeds were eliminated and aldicarb was applied, and weeds were permitted to grow and aldicarb was applied, Assumption Parish, Louisiana, 1985 and 1986; bars with different letters within the same year are significantly different ($P < 0.05$), two-way ANOVA, Tukey's HSD.

Other ant species (*e.g.*, *Pheidole dentata* Mayr and *P. floridana* Emery) are also important (Adams et al., 1981) (Table 2); combined predation by different ant species can reduce *D. saccharalis* damage, including deadhearts, by >90% (Bessin and Reagan, 1993). Exclusion of *S. invicta* and other ants from sugarcane using pesticides increases *D. saccharalis* injury (Long et al., 1958; Hensley et al., 1961; Negm and Hensley, 1967, 1969; Reagan et al., 1972; Adams et al., 1981; Showler et al., 1991).

Solenopsis invicta foraging territorial areas are substantially smaller in weedy sugarcane plots, permitting denser colonization and better *D. saccharalis* control than in weed-free sugarcane habitats (Showler et al., 1989). Showler and Reagan (1991) reported that in field plots where weeds were controlled, crop injury from *D. saccharalis* increased 2.5-fold, comparable to an application of soil-incorporated aldicarb (Figure 1). In Brazil, *Solenopsis* spp. are the most abundant ants in sugarcane, and ants were shown to suppress populations of *D. saccharalis* (Rossi and Fowler, 2000).

Table 3. Spider families collected from sugarcane fields in Assumption, Saint Martin, and West Baton Rouge, parishes, Louisiana, and Cameron, Hidalgo, and Willacy counties, Texas (Ali and Reagan, 1986; Showler and Reagan, 1991; Breene et al., 1993)

Anyphaenidae
Araneidae
Clubionidae
Dictynidae
Gnaphosidae
Linyphiidae
Lycosidae
Mysmenidae
Nesticidae
Oxyopidiae
Philodromidae
Pholcidae
Pisauridae
Salticidae
Tetragnathidae
Theridiidae
Thomisidae
Uloboridae

Spiders were collected using pitfall traps, sweep nets, aspirators, and by hand.

While weeds are competitive against sugarcane (Fontenot and Sanders, 1985: Ali and Reagan, 1985b: Ali et al., 1986: Showler and Reagan, 1991), conservation of weeds where they will not reduce crop yield might provide habitats for prey items that sustain predator populations (including *S. invicta*) (Ali et al., 1984; Showler et al., 1989; Showler and Reagan, 1991) (Table 4).

Conservation of weeds along field borders has been suggested for maintaining parasitic tachinid fly (*Lixophaga sphenopheri* [Villen.]) populations against the New Guinea sugarcane weevil (*Rhabdoscelus obscurus* Boisd.) in Hawaii (Topham and Beardsley, 1975), and because uncontrolled weeds are deleterious, perhaps limited weed habitat "islands" within sugarcane fields (Ali and Reagan, 1985a) or keeping weeds confined to furrows by conducting early season weed control only on the row tops, might be judicious tactics (Showler and Reagan, 1991).

Table 4. Relative abundances of selected prey and predator arthropods found on the soil surface, weeds, and sugarcane foliage and stalks in Assumption Parish, Louisiana, 1985 and 1986; W, most abundant (*P* < 0.05) in weedy sugarcane plots; WF, most abundant in plots where weeds were eliminated; NS, no significant differences were detected; --, not present (from Showler and Reagan, 1991)

	Soil surface[a]	Weed foliage[a]	Sugarcane foliage[a]	Sugarcane stalks[a]
Herbivorous prey	W	W	W	--
Stylet-feeding[b]	W	W	W	--
Chewing[c]	W	W	NS	--
Nonpredaceous Diptera[d]	NS	W	W	--
Predaceous arthropods[e]	W	W	W	W
Araneae[f]	W	W	--	NS
Carabidae	NS	--	--	WF
Dermaptera	NS	--	--	NS
Formicidae				
Solenopsis invicta[g]	W		NS	W
Other formicid spp.[h]	NS	W	WF	NS
Reduviidae	NS	W	NS	NS
Staphylinidae	NS	--	--	WF
Other predators[i]	W	W	NS	NS

[a] Soil surface sampled using pitfall traps, weed and sugarcane foliage sampled using sweep nets, sugarcane stalks sampled by visual inspection under leaf sheaths; data do not include numbers of *Diatraea saccharalis* found in or on the stalks.

[b] Mainly aphids, leafhoppers, seed bugs, and stink bugs.

[c] Mainly herbivorous centipedes, leaf beetles, weevils, other beetle families, and caterpillars.

[d] Flies and gnats.

[e] Araneae, spiders; Carabidae, ground beetles; Dermaptera earwigs; Formicidae, ants; Reduviidae, assassin bugs; and Staphylinidae, rove beetles..

[f] Mainly the families Aranaeidae, Clubionidae, Linyphiidae, Lycosidae, Nesticidae, Oxyopidae, Pisauridae, Salticidae, Theridiidae, and Thomisidae.

[g] Red imported fire ant.

[h] "Other formicid species were *Paratrechina vividula* and *Pheidole moerens*.

[i] "Other" predators were comprised of Cicindelidae, Coccinelidae, Elateridae, Enicocephalidae, *Geocoris* spp., and parasitic hymenoptera.

6.2. Mexican Rice Borer, *Eoreuma Loftini* (Dyar) (Lepidoptera: Crambidae)

Eoreuma loftini is native to western Mexico where it is a major pest of sugarcane (Morrill, 1925; Van Zwaluwenberg, 1926). The insect spread by the mid 1970s to eastern Mexico (Johnson, 1984), and it was first detected in the United States in 1980 by way of the Lower Rio Grande Valley of Texas (Johnson, 1984; Johnson and Van Leerdam, 1981); by 1987 it had invaded 40 counties in south Texas, by 1989 it spread into rice producing areas of east Texas (Browning et al., 1989; Reay-Jones et al., 2008), and in 2008 it was attacking rice in western Louisiana (Hummel et al., 2008, 2010). *Eoreuma loftini* has also been found in Arizona and in Imperial County, California (Johnson, 1984), but it is not economically important there. Because *E. loftini* prefers corn over other crop plants (Showler et al., 2011), its assumed range might have been underestimated, similar to the eldana borer (*Eldana saccharina* Walker) in South Africa (also found to prefer corn) (Assefa et al., 2008). This suggests that *E. loftini* might spread in the United States more quickly than anticipated (Reay-Jones et al., 2008).

The pest chiefly lays its eggs within folds of dry sugarcane leaves (Showler and Castro, 2010b). Ninety-six percent of the eggs are deposited on the basal portion of sugarcane plants, and more eggs are deposited on dead shoots than on dry leaves later in the season (Van Leerdam et al., 1986). *Eoreuma loftini*, unlike *D. saccharalis*, is considered to be a "plant stress insect," attracted to senescing foliage and to drought stressed plants for oviposition (Reagan et al., 2010; Showler and Castro, 2010a). Early instars consume leaf tissue under fresh leaf sheaths while some bore into the leaf midrib; later instars tunnel into the main stalk (Wilson, 2011). Infestations result in deadheart, lower sugar production, and stunting or lodging of stalks sometimes severe enough to make harvest unfeasible (Johnson, 1985; Legaspi et al., 1997; Hummel et al., 2008). Tunneling generally occurs in lower internodes regardless of the plant's stage of growth, and internodes are most prone to attack within 70 days of initial formation (Ring et al., 1991). The horizontal and vertical tunnels are packed with frass, blocking predators and parasitoids from entry (although the effect of *S. invicta* predation has not yet been reported) (Showler and Reagan, 2012). Pupation occurs within the stalk much in the same way as described for *D. saccharalis* (Showler and Reagan, 2012). A life cycle takes 30–45 days in subtropical south Texas, and there are 4–6 overlapping generations per year (Johnson, 1985; Legaspi et al., 1997).

Eoreuma loftini displaced *D. saccharalis* as the key sugarcane pest of south Texas (Van Leerdam et al., 1984; Legaspi et al., 1997). Because of its prevalence and its tunneling habit in the Lower Rio Grande Valley, ≈20% of sugarcane internodes are damaged and the tunnels also provide portals for red rot which breaks down sugar (Van Zwaluwenberg, 1926; Osborn and Phillips, 1946; Johnson, 1985). Percentage of bored internodes (cumulative with internode age) increased with internode age for 70–80 days, then reached a plateau at ≈55% (Ring et al., 1991). Although population densities were ≤1 larva per sugarcane stalk in south Texas (Meagher et al., 1996), as much as 50% of the internodes were bored on some varieties (Johnson, 1984). Infestations result in losses of US$575 per hectare (Meagher et al., 1994) and US$10–20 million per year (Legaspi et al., 1997, 1999), and economic consequences of *E. loftini* infestation of Louisiana are expected to include US$220 million from sugarcane injury US$45 million from damage to rice (Reay-Jones et al., 2008). In corn, *E. loftini* stalk boring and secondary infection by stalk rot pathogens can cause shattering, lodging, and complete collapse of stalks (Showler et al., 2011); by harvest, >50% of stalks in susceptible varieties are destroyed (Showler, unpublished data), but economic effects on corn production have not been well documented.

More than 17 species of graminaceous noncrop, or weed, host plants, and at least six crop host plant species can be infested by *E. loftini* (Table 2) (Showler et al., 2011). Selected noncrop host plant species were ranked, based on numbers of eggs and larval entry holes in the stalks, in the following order: sudangrass>johnsongrass and barnyardgrass>Vasey's grass>broadleaf signalgrass (Showler et al., 2011). Preference for one noncrop host plant species over another was not related to the amount of stems per plant, but it was associated with greater stem diameter and abundance of dry leaves (Showler et al., 2011). *Eoreuma loftini* host plant preferences, however, are subject to seasonal change (Beuzelin et al., 2011). Crop plant stalk diameters and quantities of dry leaf tissue, on the other hand, were not associated with numbers of eggs or larval entry holes when comparisons were made between corn, sorghum, and sugarcane (Showler et al., 2011) (Figure 2). Of the four species of major stalkborers found in corn (*Diatraea lineolata* Walker, *D. saccharalis*, *E. loftini*, and *Elasmopalpus lignoselus* [Zeller]), in the Lower Rio Grande Valley between 1982 and 1985, *E. loftini* was most prevalent (Youm et al., 1988). *Eoreuma loftini* oviposition preference on selected host crop species (on a whole plant basis) was corn>sugarcane and sorghum>rice (Reay-Jones et al., 2007; Showler et al., 2011). The ≥8.2-fold stronger oviposition preference for corn over sugarcane suggests that monitoring for the pest should occur beyond rice and sugarcane growing areas because it might have moved into corn and sorghum growing areas undetected (Showler et al., 2011). Stalkborers readily move from early-planted crops to later crops when the former are no longer suitable (Youm et al., 1988). In the Lower Rio Grande Valley, *E. loftini* was shown to move from corn when harvested in late summer to sugarcane where damage increased significantly until harvest in the late fall and winter (Showler et al., 2012). Effects of in-field weed growth on *E. loftini* has not been studied, and impacts of natural enemies have not been demonstrated largely because of the larvae's habit of blocking tunnels with frass (Legaspi et al., 1997; Showler and Reagan, 2012). Also, *S. invicta* is not as numerous in the Lower Rio Grande Valley even when weeds dominate irrigated cotton (Showler and Greenberg, 2003), but the impact of *S. invicta* might be revealed once *E. loftini* moves into Louisiana's sugarcane growing areas. Boll weevils, (*Anthonomus grandis grandis* Boheman), are also cryptic and difficult to find by natural enemies, but predation by *S. invicta* accounts for most mortality in relatively wet eastern Texas; in the Lower Rio Grande Valley of Texas, where it is too dry for *S. invicta* to thrive, the chief mortality factor is desiccation (Showler, 2007).

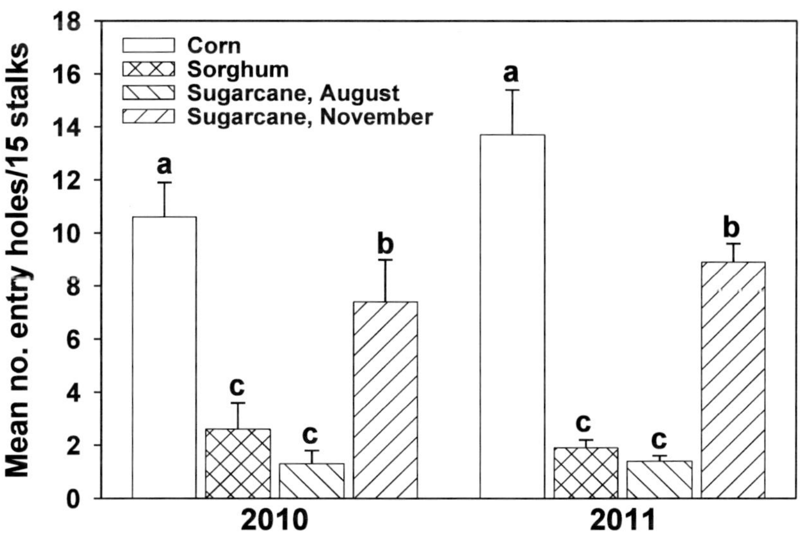

Figure 2. Mean (± SE) numbers of larval Mexican rice borer entry holes per 15 stalks of sugarcane, corn, and sorghum when the corn and sorghum were harvested (August of each year) and of sugarcane at its harvest (November of each year), field plot experiment, Hidalgo County, Texas, 2010, 2011; bars with different letters within the same year are significantly different ($P < 0.05$), two-way ANOVA, Tukey's HSD.

At this time, the main interaction with weeds, and especially corn, is the likelihood that stalkborers will move from wild hosts into sugarcane (Nagarkatti and Nair, 2001). While corn attracts *E. loftini* in greater numbers than sugarcane, corn is often harvested months earlier than sugarcane and this has been shown to increase levels of injury late in the season (Showler et al., 2012). Transgenic *Bt*-corn, however, is resistant to *E. loftini* without deterring oviposition (ATS, unpublished data); such varieties might be useful not only for protection of corn, but as a "dead end" trap crop that diverts the pest away from sugarcane.

6.3. Eldana Borer, *Eldana Saccharina* Walker (Lepidoptera: Pyralidae)

Eldana saccharina occurs mainly on *Cyperus* spp. in west sub-Saharan Africa, but during the last 50 years it has moved onto wild graminaceous hosts and sedges throughout all sub-Saharan regions, including Madagascar (Girling, 1978; Atkinson and Carnegi, 1989). In some countries, the pest has not yet been found on crop plants (*e.g.*, Ethiopia) (Assefa et al., 2006b), but *E. saccharina* is the most destructive pest of sugarcane in other parts of its distribution (Conlong, 1994; Keeping and Meyer, 2002) or it has a somewhat lesser role as part of a stalkborer complex (Conlong, 1994). A unique aspect of *E. saccharina* is that biotypes in separate parts of the insect's range have different larval feeding behaviors, host plant preferences, and natural enemies (Atachi et al., 2005; Assefa et al., 2006a, 2009).

A life cycle lasts 2–3 months and generations overlap (Girling, 1978). Clusters of eggs are deposited in crevices, such as under leaf sheaths, in leaf litter, between the stalk and the soil, and particularly within folds of dry leaf tissue of all host plants (Dick, 1945; Carnegie, 1974; Girling, 1978; Atkinson, 1980; Leslie, 1990; Mabulu and Keeping, 1999; Atachi et al., 2005). Like *E. loftini*, *E. saccharina* is also a "plant stress insect" (Showler and Reagan, 2012), preferring to lay its eggs on sugarcane stressed from lack of moisture and generally poor growing conditions (Carnegie, 1981; Moyal, 1995a; Mazodze et al., 1999). Early instars feed while exposed on leaf surfaces (but young tillers are not commonly damaged [Cochereau, 1982]), spreading on the foliage until boring into the leaf midrib (Carnegie, 1974; Girling, 1978; Kaufmann, 1983). Later instars bore into the stalk, and their tunneling produces frass that frequently extrudes from the entry holes (Carnegie, 1974; Girling, 1978). After 20–60 days (Carnegie, 1974), the last instar chews an exit hole with a thin window of host tissue over it (Girling, 1978), but some pupae are also attached to the exterior of the stalk, often beneath a leaf sheath (Carnegie, 1974). After 8–13 days as pupae, adults emerge (Carnegie, 1974).

Eldana saccharina is a moderate to very severe pest (Girling, 1972; Atkinson and Carnegie, 1989) in many sugarcane production areas within its range, and it is the chief pest of sugarcane in South Africa and Ivory Coast (Moyal, 1995a; Keeping and Meyer, 2002). Damaged stalks produce less sucrose and mass than noninjured stalks (Goebel and Way, 2003; Cochereau, 1982) reported that a 0.5% loss of sugar occurred for every 1% of internodes bored, and that up to 5.4% of sugar yield in Ivory Coast is lost to *E. saccharina*. Sugarcane losses in South Africa are valued at ≈US$10 million annually (Black et al., 1995), and infestations can become so severe that consignments are rejected at the sugar mill, followed by ratoon crop failure (Atkinson and Carnegie, 1989). Larvae prefer older stalks, but they can cause serious damage to sugarcane of all ages (Atkinson and Carnegie, 1989).

Although *E. saccharina*'s primary hosts are *Cyperus* spp. (Atkinson, 1980; Conlong and Rutherford, 2009), when graminaceous crops are cultivated near wetland areas with sedges, the pest can move to and become a pest of the crops (Atkinson, 1980; Conlong, 1997; Polaszek, 1998; Assefa et al., 2006b). Neonate *E. saccharina* show a distinct preference for native *Cyperus* hosts over sugarcane (Conlong et al., 2007) and feeding on those *Cyperus* plants produces more fecund females than those when larvae are reared on crop hosts (Atkinson and Nuss, 1989). Where *E. saccharina* was first detected in sedges near sugarcane in Zimbabwe 1987, a severe outbreak occurred in sugarcane 12 years later, then it spread into sugarcane throughout the country (Mazodze et al., 1999; Mazodze and Conlong, 2003).

Other host plants include corn, millet (*Eleusine coracana* [L.] Gaertn.), rice, sorghum, and cassava (*Manihot esculenta* Crantz) (Atkinson, 1980; Cochereau, 1982; Kaufmann, 1983; Moyal, 1995a). Ovipositioning females prefer corn over sugarcane and populations of larvae can be up to 15-fold greater than in sugarcane (Cochereau, 1982). *Eldana saccharina* decreases the grain weight of corn, and other measures of yield, in Nigeria, and facilitates infection by stalk rots (Bosque-Pérez and Mareck, 1991; Moyal, 1995b). The pest's distribution in South Africa is more widespread than sugarcane growing areas because the insect has been observed infesting indigenous grasses in major corn producing regions (Assefa et al., 2008). While *E. saccharina* is present in three indigenous wetland sedges of Ethiopia (*Cyperus papyrus* L., *Cyperus fastigiatus* Rottb., and *Cyperus dives* Delile), the insect has not been detected in graminaceous crop and noncrop plants that are commonly infested in western Africa, including elephant grass (*Pennisetum purpureum* Moench), and wild sorghum (*Sorghum bicolor arundinaceum* [Desv.] Stapf) (Girling, 1972; Conlong, 2001; Assefa et al., 2006b). Noncrop hosts, other than *Cyperus* spp., include jaraguagrass (*Hyparrhenia rufa* [Nees] Stapf), itchgrass (*Rottboellia exaltana* L.), guineagrass, wild sorghum, elephant grass, and even the broadleaf, pigweed (*Amaranthus dubius* [Mart. ex Thell.]) (Carnegie, 1974), and in Zimbabwe the pest was also found in bulrush plants (*Typha latifolia* L.) (Cochereau, 1982; Mazodze and Conlong, 2003).

7. Integrated Pest Management and Vegetational Diversity

For the most part, weeds in sugarcane systems are detrimental as competitors with the crop and as reservoirs for diseases, arthropod pests, and vectors of sugarcane diseases. The extent to which weeds are problematic for stalkborer pests affecting sugarcane has not been clearly assessed, but weed hosts are probably important reservoirs, particularly during winter. The presence of one or more alternate crop hosts of *E. loftini* in an agricultural landscape can contribute to greater infestations of sugarcane by the pest unless strategies are developed on an area-wide, as well as on a field, basis to avert it. While elimination of host weeds year-round within and outside the field is likely useful by reducing breeding and food resources, planting resistant "dead end" trap corn varieties, and modifying alternate host crop harvest times to reduce movement into sugarcane (and other remaining host crops of *E. loftini*), might prove useful as pest management tactics.

The area-wide landscape is also important to *E. saccharina*, involving proximity of wetlands and sedge hosts, and some graminaceous and broadleaf weed species. The chief concern is the ability of *E. saccharina* to shift from wetland sedges to sugarcane and other crops. Corn might be problematic in the same way as it might be for *E. loftini*, and integrated pest management tactics similar to those suggested for *E. loftini* could be useful against *E. saccharina*. Although graminaceous weeds have not been evaluated as in-field alternate hosts of *E. saccharina*, it is likely that weed control would offer the added benefit of eliminating possible reservoirs.

Weeds have been shown to be associated with reduced numbers of some phytophagous nematode species, but even where root knot nematodes (*Meloidogyne incognita* [Kofoid and White] Chitwood and *M. javanica*

[Treub] Chitwood), were augmented in the soil, populations were relatively low in weedy habitats (Showler et al., 1991). This possible benefit of weeds, however, is masked by their deleterious competitive effects (Showler and Reagan, 1991).

Weeds can have a positive role in enhancing natural enemy populations against *D. saccharalis*, and this is especially true in wet regions where *S. invicta* thrives. When *E. loftini* spreads into sugarcane growing areas of Louisiana with large *S. invicta* populations, it is possible that the ant will suppress it as well, but in the drier Lower Rio Grande Valley of Texas, natural enemies have had negligible impact (Showler and Reagan, 2012) and it seems unlikely that permitting weeds in the field will substantially improve predator activity against the pest. Where *S. invicta* foraging is sufficient to reduce stalkborers, however, the problem of weed competition must still be ameliorated. Research has not yet been conducted on how weeds might be permitted in or around sugarcane fields that minimizes or eliminates the problems associated with competition while encouraging colonization by *S. invicta* and other natural enemies.

REFERENCES

Abbott E.V. and Charpentier L.J. (1963) Additional insect vectors of sugarcane mosaic. *Proceedings of the International Society of Sugarcane Technologists* 11: 755-760.

Abbott E.V. and Tippett R.L. (1966) Strains of sugarcane mosaic virus. USDA Bulletin 1340, Washington, DC, USA.

Adams C.T., Summers T.E., Lofgren C.S., Focks D.A. and Prewitt J.C. (1981) Interrelationship of ants and the sugarcane borer in Florida sugarcane fields. *Environmental Entomology* 10: 415-418.

Akbar W., Showler, A.T., Reagan T.E. and White W.H. (2010) Categorizing sugarcane cultivar resistance to the sugarcane aphid and yellow sugarcane aphid (Hemiptera: Aphididae). *Journal of Economic Entomology* 103: 1431-1437.

Ali A.D., Reagan T E. and Flynn J.L. (1984) Influence of selected weedy and weed-free sugarcane habitats on diet composition and foraging activity of the imported fire ant (Hymenopera: Formicidae). *Environmental Entomology* 13: 1037-1041.

Ali A.D. and Reagan T.E. (1985a) Vegetation manipulation impact on predator and prey populations in Louisiana sugarcane ecosystems. *Journal of Economic Entomology* 78: 1409-1414.

Ali A.D. and Reagan T.E. (1985b) Spider inhabitants of sugarcane ecosystems in Louisiana: an update. *Proceedings of the Louisiana Academy of Science* 48: 18-22.

Ali A.D. and Reagan T.E. (1986) Influence of selected weed control practices on araneid faunal composition and abundance in sugarcane. *Environmental Entomology* 15: 527-531.

Ali A.D., Reagan T.E. and Flynn J.L. (1986) Effects of johnsongrass (*Sorghum halepense*) density on sugarcane (*Saccharum officinarum*) yield. *Weed Science* 14: 381-383.

Allen W.W. and Smith R.F (1958) Some factors influencing the efficiency of *Apanteles mediacaginis* Muesebeck (Hymenoptera: Braconidae) as a parasite of the alfalfa caterpillar, *Colia philodice eurytheme* Boisduval. *Hilgaria* 28: 1-43.

Altieri M.A.; Van Schoonhoven, A. and Doll, J.D. (1977) The ecological role of weeds in insect pest management systems: a review illustrated b y bean (*Phaseolus vulgaris*) cropping systems. *Tropical Pest Management (PANS)* 23: 195-205.

Altieri M.A., Francis C.A., Van Schoonhoven A. and Doll J.D. (1978) A review of insect prevalence in maize (*Zea mays* L.) and bean (*Phaseolus vulgaris* L.) polycultural systems. *Field Crops Research* 1: 33-49.

Altieri, M.A. and Whitcomb, W.H. (1979) The potential use of weeds in the manipulation of beneficial insects. *Hortscience* 14: 12-18, ISSN 0018-5345.

Altieri M.A. and Whitcomb W.H. (1980) Weed manipulation for insect pest management in corn. *Environmental Entomology* 4: 483-489, ISSN 0046-225X.

Altieri M.A. (1981) Weeds may augment biological control of insects. *California Agriculture* 35: 22-24.

Altieri M.A. and Todd, J.W. (1981) Some influences of vegetational diversity on insect communities of Georgia soybean fields. *Journal of Environmental Protection and Ecology* 3: 333-338.

Altieri M.A. (1984) Patterns of insect diversity in monocultures and polycultures of brussels sprouts. *Journal of Environmental Protection and Ecology* 6: 227-232.

Altieri M.A. and Schmidt L.L. (1986) Cover crops affect insect and spider populations in apple orchards. *California Agriculture* 40: 15-17.

Arevelo R.A., Cerrizuela E.A. and Olea I.L. (1977) Competition from specific weeds in sugarcane II. *Sorghum halepense* (L.) Pers. *Revue Agronomy Noroeste Argentina* 14: 39-51.

Assefa Y., Mitchell A. and Conlong D.E. (2006a) Phylogeography of *Eldana saccharina* Walker (Lepidoptera: Pyralidae). *Annals of the Society of Entomology Francaise* 42: 331-337.

Assefa Y., Conlong D.E. and Mitchell, A. (2006b) Status of *Eldana saccharina* (Lepidoptera: Pyralidae), its host plants and natural enemies in Ethiopia. *Bulletin of Entomological Research* 96: 497-504.

Assefa, Y.; Conlong, D.E.; Van den Berg, J. and Le Rü, B.P. (2008). The wider distribution of *Eldana saccharina* (Lepidoptera: Pyralidae) in South Africa and its potential risk to maize production. *Proceedings of the South African Sugarcane Technologists Association* 81: 290-297, ISSN 03701816.

Assefa Y., Van den Berg J., Mitchell A., Le Rü B.P. and Conlong D.E. (2009) Record of *Eldana saccharina* Walker (Lep., Pyralidae) in inland South Africa and its genetic relationship with the coastal population. *Journal of Applied Entomology* 133: 449-455.

Astudillo G.E. and Birchfield W. (1980) Pathology of *Hoplolaimus columbus* on sugarcane. *Phytopathology* 70: 565.

Atachi P., Sekloka E.T. and Schulthess F. (2005) Study on some bioecological aspects of *Eldana saccharina* Walker (Lep., Pyralidae) on *Zea mays* L. and alternative host plants. *Journal of Applied Entomology* 129: 447-455.

Atkinson P.R. (1980) On the biology, distribution and natural host plants of *Eldana saccharina* Walker. *Journal of the Entomological Society of South Africa* 43: 171-194.

Atkinson P.R. and Carnegie A.J.M. (1989) Population dynamics of the sugarcane borer, *Eldana saccharina* Walker, in Natal, South Africa. *Bulletin of Entomological Research* 79: 61-80.

Atkinson P.R. and Nuss K.J. (1989) Associations between host-plant nitrogen and infestations of the sugarcane borer, *Eldana saccharina* Walker (Lepidoptera: Pyralidae). *Bulletin of Entomological Research* 79: 489-506.

Atsatt P.R. and O'Dowd, D.J. (1976) Plant defense guilds. *Science* 193: 24-29.

Bach C.E. (1980) Effect of plant diversity and time of colonization on an herbivore-plant interaction. *Oecologia* 44: 319-326.

Barney R.J., Lamp W.O., Armbrust E.J., and Kapusta G. (1984) Insect predator community and its response to weed management in spring-planted alfalfa. *Environmental Protection and Ecology* 6: 23-33.

Benda G.T.A. (1969) Sugarcane mosaic virus in Louisiana: some aspects of a chronic problem. *American Society of Sugarcane Technologists* 16: 61-81.

Bendixon L.E., Reynolds D.A. and Riedel R.M. (1979) An annotated bibliography of weeds as reservoirs for organisms affecting crops I. Nematodes. *Ohio Agricultural Research and Development Center, Research Bulletin* 1109.

Bessin R.T. and Reagan T.E. (1993) Cultivar resistance and arthropod predation of sugarcane borer (Lepidoptera: Pyralidae) affects incidence of deadhearts in Louisiana sugarcane. *Journal of Economic Entomology* 86: 929-932.

Beuzelin J.M., Reagan T.E., Meszaros A., Wilson L.T., Way M.O., Blouin D. and Showler A.T. (2011) Seasonal infestations of two stem borers (Lepidoptera: Crambidae) in non-crop grasses of Gulf Coast rice agroecosystems. *Environmental Entomology* 40: 1036-1050.

Birchfield W. (1972) Nematicides on sugarcane. Report to the Sugarcane Contact Commission and Louisiana Agricultural Experiment Station, pp. 71-73.

Birchfield W. (1984) Nematode parasites of sugarcane. In *Plant and Insect Nematodes*, Nickel, W.R. (ed.), pp. 571-587. Marcel Dekker, ISBN 9780824770792, New York, New York, USA.

Blackburn F. (1984) *Sugar Cane*. Longman, ISBN 9780470204184, New York, New York, USA.

Black K.G., Huckett B.I. and Botha F.C. (1995) Ability of *Pseudomonas florescence* engineered for insecticidal activity against sugarcane stalk borer to colonise the surface of sugarcane plants. *Proceedings of the South African Sugar Technologists Association* 69: 21-24.

Bosque-Pérez N.A. and Mareck J.H. (1991) Effect of the stem borer *Eldana saccharina* (Lepidoptera: Pyralidae) on the yield of maize. *Bulletin of Entomological Research* 81: 243-247.

Brandenburg R.L. and Kennedy G.C. (1981) Overwintering of the pathogen *Entomophthora floridana* and its host, the two spotted spider mite. *Journal of Economic Entomology* 74: 428-431.

Brandes E.W. (1920) Artificial and insect transmission of sugarcane mosaic. *Journal of Agricultural Research* 19: 131-138.

Breene R.G., Meagher R.L. and Dean D.A. (1993) Spiders (Araneae) and ants (Hymenoptera: Formicidae) in Texas sugarcane fields. *Florida Entomologist* 76: 645-650, ISSN 0015-4040.

Browning H.W. and Melton C.W. (1987) Indigenous and exotic trichogrammatids (Hymenoptera: Trichogrammatidae) evaluated for biological control of *Eoreuma loftini* and *Diatraea saccharalis* (Lepidoptera: Pyralidae) borers on sugarcane. *Environmental Entomology* 16: 360-364.

Browning H.W., Way M.O. and Drees B.M. (1989) Managing the Mexican rice borer in Texas. Texas Agricultural Extension Service Bulletin B-1620.

Capinera J.L. (2001) Sugarcane borer, *Diatraea saccharalis* (Fabricius) (Insecta: Lepidoptera: Pyralidae). Entomology and Nematology Department, Florida Cooperative Extension Service, University of Florida, Gainesville, Florida, USA.

Carnegie A.J.M. (1974) A recrudescence of the borer *Eldana saccharina* Walker (Lepidoptera: Pyralidae). *Proceedings of the South African Sugarcane Technologists Association* 41: 107-110.

Carnegie A.J.M. (1981) Combating *Eldana saccharina* Walker, a progress report. *Proceedings of the South African Sugarcane Technologists Association* 48: 107-110.

Carter C.E. (1977) Excess water decreases cane and sugar yields. *Journal of the American Society of Sugar Cane Technologists* 6: 44-51.

Charpentier L.J., McCormick W.J. and Mathes R. (1967) Beneficial arthropods inhabiting sugarcane fields and their effects on borer infestations. *Sugar Bulletin* 45: 276-277.

Cochereau P. (1982) Observations on the borer *Eldana saccharina* Walker (Lep., Pyralidae) in maize and sugarcane in Ivory Coast. *Proceedings of the South African Sugarcane Technologists Association* 49: 82-84.

Coll M. and Bottrell D. G. (1995) Predator-prey association in mono- and dicultures: effect of maize and bean vegetation. *Agriculture, Ecosystems and Environment* 54: 115-125.

Conlong D.E. (1994) A review and perspectives for the biological control of the African sugarcane stalkborer *Eldana saccharina* Walker (Lepidoptera: Pyralidae). *Agriculture, Ecosystems and Environment* 48: 9-17.

Conlong D.E. (1997) Impact of ecology and host plant shifts of *Eldana saccharina* (Lepidoptera: Pyralidae) on its biological control in graminaceous crops. *Proceedings of the South African Sugarcane Technologists Association* 71: 94.

Conlong D.E. (2001) Biological control of indigenous African stemborers: what do we know? *Insect Science and Its Applications* 21: 212-213.

Conlong D.E., Karl B. and Byrne M. (2007) Independent kids or motherly moms? Implications for integrated pest management of *Eldana saccharina*

Walker (Lepidoptera: Pyralidae). *Proceedings of the International Society of Sugar Cane Technologists* 26: 787-796.

Conlong D.E. and Rutherford R.S. (2009) Conventional and news biological and habitat interventions for integrated pest management systems: review and case studies using *Eldana saccharina* Walker (Lepidoptera: Pyralidae), pp. 241-261. In *Integrated Pest Management: Innovation-Development Process*, Peshin, R. and Dhawan, A.K. (eds.), Springer, ISBN 9781402089916, New York, New York, USA.

Dempster J.P. (1969) Some effects of weed control on the numbers of the small cabbage white (*Pieris rapae* L.) on brussels sprouts. *Journal of Applied Ecology* 6: 339-405.

Denmark H.A. (1988) Sugarcane aphids in Florida. Florida Department of Agriculture and Consumer Services, Division of Plant Industry, Entomology Circular 302.

Dick J. (1945) Some data on the biology of the sugarcane borer (*Eldana saccharina*, Wlk.) *Proceedings of the South African Sugarcane Technologists Association* 19: 75-79.

Dissayanake N., Hoy J.W. and Griffin J.L. (1997) Weed hosts of the sugarcane root pathogen, *Pythium arrhendomanes*. *Plant Disease* 81: 587-591.

Durigan J.C. (2005) Effects of plant densities and management of purple nutsedge on sugarcane yield and effect on growth stages and main way of herbicides contact and absorption on the control of tubers. *Journal of Environmental Science and Health* 40: 111-117.

Edgerton C.W. (1939) Stubble deterioration. *Proceedings of the International Society of Sugar Cane Technologists* 6: 334-341, ISSN 13902962.

Edgerton C.W. (1959) Sugarcane and its diseases. Louisiana State University Press, ISBN 2876143860, Baton Rouge, Louisiana, USA.

FAO (Food and Agriculture Organization of the United Nations) (2009) Major food and agricultureal commodies and producers. http://www.fao.org/ess/top/commodity.html? lang+enanditem=156andyear=2005.

Fisher B.B., Burtch L.M., and Smith R. (1977) The role of selective herbicides in sugar beet production. U.C. Cooperative Extension, Fresno County, California, USA.

Fontenot D.B. and Sanders D. (1985) Louisiana guide to controlling johnsongrass seedlings and annual weeds in sugarcane – spring 1985. Louisiana State University Cooperative Extension Service Publication, Baton Rouge, Louisiana, USA.

Forbes I.L., Giamalva M. and Falgout B. (1967) Mosaic found in johnsongrass in the Louisiana sugarcane area. *Sugar Bulletin* 45: 278.

Foster M.A. and Ruesink W.G. (1984) Influence of flowering weeds associated with reduced tillage in corn on a black cutworm (Lepidoptera: Noctuidae) parasitoid, *Meteorus rubens* (Nees von Esenbeck). *Environmental Entomology* 13: 664-668.

Fuchs T.W., Harding J.A. and Smith J.W. (1979) Induction and termination of diapause in the sugarcane borer. *Annals of the Entomological Society of America* 72: 271-274.

Fuller B.W. and Reagan T.E. (1989) The relationship of sweet sorghum plant fiber and survival of the sugarcane borer, *Diatraea saccharalis* (F.) (Lepidoptera: Pyralidae). *Journal of Agricultural Entomology* 6: 113-118.

Gaskin T.A. (1958) Weed hosts of *Meloidogyne incognita* in Indiana. *Plant Disease Reporter* 42: 802-803.

Gibson I.A.S. and Jones T. (1977) Monoculture as the origin of major forest pests and diseases, especially in the tropics and southern hemisphere. In: *Origins of Pest, Parasite, and Weed Problems*, J.M. Cherrett and G.R. Sagar (eds.), pp. 139-161. Blackwell Science, ISBN 0-632-00226-3, Oxford, England.

Gillespie A.G. and Koike H. (1973) Sugarcane mosaic virus and maize dwarf virus in Mexico infections of sugarcane and other grasses. *Phytopathology* 63: 1300-1307.

Girling D.J. (1972) *Eldana saccharina* Wlk. (Lepidoptera: Pyralidae), a pest of sugarcane in East Africa. *Proceedings of the International Society of Sugarcane Technologists* 14: 429-434.

Girling D.J. (1978) The distribution and biology of *Eldana saccharina* Walker (Lepidoptera: Pyralidae) and its relationship to other stem-borers in Uganda. *Bulletin of Entomological Research* 68: 471-488.

GMO Compass (2008) Sugarcane. http://www.gmo-compass.org/eng/ database/plants/ 76.sugarcane.html.

Goebel F.R. and Way M.J. (2003) Investigation of the impact of *Eldana saccharina* (Lepidoptera: Pyralidae) on sugarcane yield in field trials in Zululand. *Proceedings of the South African Sugarcane Technologists Association* 77: 256-64.

Golden A.M. and Birchfield W. (1965) *Meloidogyne graminicola* (Heteroderidae) a new species of root kinot nematode from grass. *Proceedings of the Helminthological Society of Washington* 32: 228-231.

Graham S.A. (1956) Forest insects and the law of natural compensation. *Canadian Entomology* 88: 45-55.

Harper J.L. (1969) The role of predation in vegetational diversity: diversity and stability in ecological systems. In *Brookhaven Symposium on Biology* pp. 22: 48-61.

Harper J.L. (1977) *Population Biology of Plants.* Academic Press, ISBN 9780824770792, New York, New York, USA.

Hatfield J.L. and Stewart B.A. (1997) *Integrated Weed and Soil Management.* CRC Press, ISBN 1575040697, Boca Raton, Florida, USA.

Hatley C.L. and McMahon J.A. (1980) Spider community organization: seasonal variation and the role of vegetation architecture. *Environmental Entomology* 9: 632-639.

Hensley S.D., Long W.H., Roddy L.R., McCormick W.J. and Cancienne E.J. (1961) Effects of insecticides on the predaceous arthropod fauna of Louisiana sugarcane fields. *Journal of Economic Entomology* 54: 146-149.

Hensley S.D. (1971) Management of sugarcane borer populations in Louisiana – a decade of change. *Entomophaga* 16: 133-146.

Hogger C.H. and Bird G.W. (1974) Weeds and covercrops as overwintering hosts of plant-parasitic nematodes of soybeans and cotton in Georgia. *Journal of Nematology* 6: 142-143.

House G.J. (1989) Soil arthropods from weed and crop roots of an agroecosystem in a wheat-soybean-corn rotation: impact of tillage and herbicides. *Agriculture, Ecosystems and Environment* 25: 233-244.

Hoy J.W. and Schneider R.W. (1988a) Role of *Pythium* in sugarcane stubble decline: effects on plant growth in field soil. *Phytopathology* 78: 1692-1696.

Hoy J.W. and Schneider R.W. (1988b) Role of *Pythium* in sugarcane stubble decline: pathogenicity and virulence of *Pythium* species. *Phytopathology* 78: 1688-1692.

Hummel N., Reagan T.E., Pollet D., Akbar W., Beuzelin J.M., Carlton C., Saichuk J., Hardy T. and Way M.O. (2008) Mexican rice borer, *Eoreuma loftini* (Dyar). Louisiana State University AgCenter Publication 3098, Baton Rouge, Louisiana, USA.

Hummel N.A., Hardy T., Reagan T.E., Pollet D., Carlton C., Stout M.J., Beuzelin J.M., Akbar W. and White W.H. (2010) Monitoring and first discovery of the Mexican rice borer, *Eoreuma loftini* (Lepidoptera: Crambidae) in Louisiana. *Florida Entomologist* 93: 123-124.

Ingram J.W. and Summers E.M. (1936) Transmission of sugarcane mosaic by the rusty plum aphid *Hysteroneura setaria* (Thos.). *Journal of Agricultural Research* 52: 879-887.

Ingram J.W. and Summers E.M. (1938) Transmission of sugarcane mosaic by the greenbug, *Toxoptera graminae* (Rond.). *Journal of Agricultural Research* 56: 537-540.

Ingram J.W., Bynum E.K., Mathes R., Haley W.E. and Charpentier W.E. (1951) Pests of sugarcane and their control. U.S. Department of Agriculture, circular 878, Washington, DC, USA.

Johnson K.J.R. and Van Leerdam M.B. (1981) Range extension of *Acigona loftini* into the Lower Rio Grande Valley of Texas. *Sugar y Azucar* 76: 119.

Johnson K.J.R. (1984) Identification of *Eoreuma loftini* (Dyar) (Lepidoptera: Pyralidae) in Texas, 1980: forerunner for other sugarcane boring pest immigrants from Mexico? *Bulletin of the Entomological Society of America* 30: 47-52.

Johnson K.J.R. (1985) Seasonal occurrence and insecticidal suppression of *Eoreuma loftini* (Lepidoptera: Pyralidae) in sugarcane. *Journal of Economic Entomology* 78: 960-966.

Kaufmann T. (1983) Behavioral biology, feeding habits, and ecology of three species of maize stem-borers: *Eldana saccharina* (Lepidoptera: Pyralidae), *Sesamia calamistis* and *Busseola fusca* (Noctuidae) in Ibadan, Nigeria, West Africa. *Journal of the Georgia Entomological Society* 18: 255-259.

Keeping M.G. and Meyer J.H. (2002) Calcium silicate enhances resistance of sugarcane to the African stalk borer *Eldana saccharina* Walker (Lepidoptera: Pyralidae). *Agricultural and Forest Entomology* 4: 265-274.

Komblas K.N. (1964) Field studies of aphid vectors of sugarcane mosaic virus and methods of control. Ph.D. dissertation, Louisiana State University, Baton Rouge, Louisiana, USA.

Laster M.L. (1974) Increasing natural enemy resources through crop rotation and strip cropping. *Proceedings of the Summer Institute of Biocontrol of Plant Insects and Disease*, pp. 137-143.

Legaspi J.C., Legaspi B.C., Irvine J.E. and Saldana R.R. (1997) Mexican rice borer, *Eoreuma loftini* (Lepidoptera: Pyralidae) in the Lower Rio Grande Valley of Texas: its history and control. *Subtropical Plant Science* 49: 53-64.

Legaspi J.C., Legaspi B. C., Irvine J.E., Meagher R.L. and Rozeff N. (1999) Stalkborer damage on yield and quality of sugarcane in the Lower Rio Grande Valley of Texas. *Journal of Economic Entomology* 92: 228-234.

Leslie G.W. (1990) The influence of dead leaf material on the oviposition behavior of *Eldana saccharina* (Lepidoptera: Pyralidae) in sugarcane.

Proceedings of the South African Sugarcane Technologists Association 64: 100-10.

Letourneau D.K. (1986) Associational resistance in squash monocultures and polycultures in tropical Mexico. *Environmental Entomology* 15: 285-292.

Long C.S., Cancienne E.A., Cancienne E.J., Dobson R.W. and Newsom L.D. (1958) Fire ant eradication program increases damage by the sugarcane borer. *Sugar Bulletin* 37: 62-63.

Luckman W.H. and Metcalf R.L. (1975) The pest management concept. In *Introduction to Pest Management*, Metcalf, R.L. and Luckman, W.H. (eds.), pp. 1-21 Wiley and Sons, ISBN 9780471085478, New York, New York, USA.

Lv J., Wilson L.T.and Longnecker M.T. (2008) Tolerance and compensatory response of rice to sugarcane borer (Lepidoptera: Crambidae) injury. *Environmental Entomology* 37: 796-807.

Mabulu L.Y. and Keeping,M.G. (1999) Glasshouse tests for ovipositional antixenosis of South African sugarcane varieties to *Eldana saccharina* (Lepidoptera: Pyralidae). *Proceedings of the South African Sugarcane Technologists Association* 73: 104-106.

Manuel J.S., Bendixon L.E., Reynolds D.A. and Riedel R.M. (1980) Weeds as hosts of *Pratylenchus. Ohio Agricultural Research and Development Center Research Bulletin* 1123.

Marsh J. (2006) Sugar – the political economy of a product. *Journal of Science, Food and Agriculture* 86: 2516-2522.

Mazodze R., Nyanthete C. and Chidoma S. (1999) First outbreak of *Eldana saccharina* (Lepidoptera: Pyralidae) in sugarcane in the south-east lowland of Zimbabwe. *Proceedings of the South African Sugarcane Technologists Association* 73: 107-111.

Mazodze R. and Conlong D.E. (2003) *Eldana saccharina* (Lepidoptera: Pyralidae) in sugarcane (*Saccharum* hybrids), sedge (*Cyperus digitatus*) and bulrush (*Typha latifolia*) in south-eastern Zimbabwe. *Proceedings of the South African Sugarcane Technologists Association* 77: 266-274.

McWhorter C.G. (1972) Factors affecting johnsongrass rhizome production and germination. *Weed Science* 20: 41-45.

McWhorter S.G. and Anderson J.M. (1981) The technical and economic effects of johnsongrass (*Sorghum halepense*) control in soy beans (*Glycine max*). *Weed Science* 29: 245-253.

Meagher R.L., Smith J.W. and Johnson K.J.R. (1994) Insecticidal management of *Eoreuma loftini* (Lepidoptera: Pyralidae) on Texas sugarcane: a critical review. *Journal of Economic Entomology* 87: 1332-1344.

Meagher R.L., Wilson L.T. and Pfannenstiel R.S. (1996) Sampling *Eoreuma loftini* (Lepidoptera: Pyralidae) on Texas sugarcane. *Environmental Entomology* 25: 7-16.

Moré M., Trumper E.V. and Prola M.J. (2003) Influence of corn, *Zea mays*, phenological stages in *Diatraea saccharalis* F. (Lep. Crambidae) oviposition. *Journal of Applied Entomology* 127: 512-515.

Morrill A.W. (1925) Commercial entomology on the west coast of Mexico. *Journal of Economic Entomology* 18: 707-716.

Moyal P. (1995a) Borer infestation and damage in relation to maize stand density and water stress in the Ivory Coast. *International Journal of Pest Management* 41: 114-121.

Moyal P. (1995b) Stemborer larval growth and tunneling and its influence on maize vegetative growth in Côte d'Ivoire. *Insect Science and Its Applications* 16: 293-301.

Nagarkatti S. and Nair R. (2001) The influence of wild and cultivated *Gramineae* and *Cyperaceae* on populations of sugarcane borers and their parasites in North India. *Entomophaga* 18: 419-430.

Nass L.L., Pereira P.A.A. and Ellis D. (2007) Biofuels in Brazil: an overview. *Crop Science* 47: 2228-2237.

Negm A.A. and Hensley S.D. (1967) The relationship of arthropod predators to crop damage inflected by the sugarcane borer. *Journal of Economic Entomology* 60: 1503-1506.

Negm A.A. and Hensley S.D. (1969) Evaluation of certain biological control agents of the sugarcane borer in Louisiana. *Journal of Economic Entomology* 62: 1003-1013.

Odum E.P. (1971) *Fundamentals of Ecology*. Saunders, ISBN 9780534420666, Philadelphia, Pennsylvania, USA.

Ogunwolu E.O., Reagan T.E. and Damman K.E. (1987) Efficacy of pest control strategies in Louisiana sugar cane: a preliminary survey. *Journal of Agricultural Science Cambridge* 108: 661-665.

Ogunwolu E.O., Reagan T.E., Flynn J.L. and Hensley S.D. (1991) Effects of *Diatraea saccharalis* (F.) (Lepidoptera: Pyralidae) damage and stalk rot fungi on sugarcane yield in Louisiana. *Crop Protection* 10: 57-61.

Oloumi-Sadeghi H., Zavaleta L.R., Lamp W.D., Armbrust E.J. and Kapusta G. (1987) Interactions of the potato leafhopper (Homoptera: Cicadellidae) with weeds in an alfalfa ecosystem. *Environmental Entomology* 16: 1175-1180.

Osborn H.T. and Phillips G.R. (1946) *Chilo loftini* in California, Arizona, and Mexico. *Journal of Economic Entomology* 39: 755-759.

Outlaw J. L., Ribera L.A., Richardson J.W., da Silva J., Bryant H. and Kloss S.L. (2007) Economics of sugar-based ethanol production and related policy issues. *Journal of Agricultural and Applied Economics* 39: 357-363.

Paray N.B., Khoodoo M.H.R., Saumtally A.S. and Ganeshan S. (2011) Vector-virus relationship for *Melanaphis sacchari* (Zehnt.) (Hemiptera: Aphididade) transmitting sugarcane yellow leaf luteovirus in Mauritius. *Sugar Technology* 13: 77-80.

Peng S.Y. (1984) *The Biology and Control of Weeds in Sugarcane*. Elsevier, ISBN 9780444421333, Amsterdam.

Pimentel D. (1961) Species diversity and insect population outbreaks. *Annals of the Entomological Society of America* 54: 76-85.

Polaszek A. (1998) *African Cereal Stem Borers: Economic Importance, Taxonomy, Natural Enemies and Control*. CAB International, ISBN 0851991750, Wallingford, Oxon, England.

Pollet, D.K., Reagan, T.E., White, W.H. and Rester, D.C. (1978). Pest management of sugarcane insects. Louisiana Cooperative Extension Publication no. 1982.

Ramirez A. (1981) Pathogenic evaluation of the nematode *Tylenchorhynchus annulatus* on sugarcane in Costa Rica. *Inter-American Sugarcane Seminar: Insect and Rodent Pests* 2: 159-164.

Rands R.D. and Dopp E. (1938) *Pythium* rot of sugarcane. United States Department of Agriculture Technical Bulletin 666.

Reagan T.E., Coburn C. and Hensley S.D. (1972) Effects of Mirex on the arthropod fauna of a Louisiana sugarcane field. *Environmental Entomology* 1: 588-591.

Reagan T.E. and Martin F.A. (1989) Breeding for resistance to *Diatraea saccharalis* (F.). In *Sugarcane Varietal Improvement*, Naidu, K.M., Sreenivasan, T.V. Premachandran, M.N. (eds.), pp. 313-331. Sugarcane Breeding Institute, ISBN 3527328408, Coimbatore, India.

Reagan T.E., Akbar W. and Showler A.T. (2010) A nutritional perspective of sugarcane resistance to stalk borers and sap feeders. *International Society of Sugar Cane Technologists* 27: 792-798.

Reay-Jones F.P.F., Wilson L.T., Showler A.T., Reagan T.E. and Way M.O. (2007) Role of oviposition on preference in an invasive crambid impacting two graminaceous host crops. *Environmental Entomology* 36: 938-951.

Reay-Jones F.P.F., Wilson L.T., Reagan T.E., Legendre B.L. and Way M.O. (2008) Predicting economic losses from the continued spread of the Mexican rice borer (Lepidoptera: Crambidae). *Journal of Economic Entomology* 101: 237-250.

Reynolds H.T., Adkisson P.L. and Smith R.F. (1975) Cotton pest management. In *Introduction to Pest Management*, Metcalf, R.L. and Luckman, W.H. (eds.), pp. 379-443. Wiley and Sons, ISBN 9780471085478, New York, New York, USA.

Richard E.P. and Dalley C.D. (2007) Sugarcane response to bermudagrass interference. *Weed Technology* 21: 941-946.

Riechert S.E. and Lockley T. (1984) Spiders as biocontrol agents. *Annual Review of Entomology* 29: 299-320.

Ring D.R., Browning H.W., Johnson K.J.R., Smith J.W. and Gates C.E. (1991) Age-specific susceptibility of sugarcane internodes to attack by the Mexican rice borer (Lepidoptera: Pyralidae). *Journal of Economic Entomology* 84: 1001-1009.

Risch S.J. (1979) A comparison, by sweep sampling, of the insect fauna from corn and sweet potato monocultures and dicultures in Costa Rica. *Oecologia* 42: 195-211.

Risch S.J. (1980) Fewer beetle pests on beans and cowpeas interplanted with banana in Costa Rica. *Turrialba* 30: 229-230.

Rodriguez-del-Bosque L.A., Smith J.W. and Martinez A.J. (1995) Winter mortality and spring emergence of corn stalkborers (Lepidoptera: Pyralidae) in subtropical Mexico. *Journal of Economic Entomology* 88: 628-634.

Rossi M.N. and Fowler H.G. (2004) Spatial and temporal population interaction between the parasitoids *Cotesia flavipes* and Tachinidae flies: considerations on the adverse effects of biological control practice. *Journal of Applied Entomology* 128: 112-119.

Ryan J., Ryan M.F. and McNaeidhe F.M. (1980) The effect of interrow plant cover on populations of the cabbage root fly, *Delia brassicae* (Wiedemann). *Journal of Applied Ecology* 17: 31-40.

Salassi M.E. (2008) Economics of sugarcane production: what does it take for this industry to survive? *Louisiana Agriculture* 51: 20-24.

Scaglusi S.M. and Lockhart B.E.L. (2000) Transmission, characterization, and serology of a Luteovirus associated with yellow leaf syndrome of sugarcane. *Phytopathology* 90: 120-124.

Sharpe P. (1998) *Sugar Cane: Past and Present*. Southern Illinois University, ISBN 0-822-31866-2, Carbondale, Illinois, USA.

Showler A.T. (1987) Interactions of weed, nematode and arthropod management strategies with the Louisiana sugarcane agroecosystem: a holistic perspective. Ph.D. dissertation, Louisiana State University, Baton Rouge, Louisiana, USA.

Showler A.T., Knaus R.M. and Reagan T.E. (1989) Foraging territoriality of the imported fire ant, *Solenopsis invicta* Buren, in sugarcane as determined by neutron activation analysis. *Insectes Sociaux* 36: 235-239.

Showler A.T., Reagan T.E. and Shao K.P. (1990) Nematode interactions with weeds and sugarcane mosaic virus in Louisiana sugarcane. *Journal of Nematology* 22: 31-38.

Showler A.T. and Reagan T.E. (1991) Effects of sugarcane borer, weed, and nematode control strategies in Louisiana sugarcane. *Environmental Entomology* 20: 358-370.

Showler A.T., Reagan T.E. and Flynn J.L. (1991) Augmentation and aldicarb treatment of nematodes in selected sugarcane weed habitats. *Journal of Nematology* 23: 717-723.

Showler A.T. (2002) Effects of water deficit stress, shade, weed competition, and kaolin particle film on selected foliar free amino acid accumulations in cotton, *Gossypium hirsutum* L. *Journal of Chemical Ecology* 28: 631-651, ISSN 0098-0331.

Showler A.T. and Greenberg S.M. (2003) Effects of weeds on selected arthropod herbivore and natural enemy populations, and on cotton growth and yield. *Environmental Entomology* 32: 39-50.

Showler A.T. (2007) Subtropical boll weevil ecology. *American Entomologist* 53: 240-249.

Showler A.T. and Castro B.A. (2010a) Influence of drought stress on Mexican rice borer (Lepidoptera: Crambidae) oviposition preference in sugarcane. *Crop Protection* 28: 722-727.

Showler A.T. and Castro B.A. (2010b) Mexican rice borer (Lepidoptera: Crambidae) oviposition site selection stimuli on sugarcane, and potential field applications. *Journal of Economic Entomology* 103: 1180-1186.

Showler A.T., Beuzelin J.M. and Reagan T.E. (2011) Alternate crop and weed host plant oviposition preferences by the Mexican rice borer (Lepidoptera: Crambidae). *Crop Protection* 30: 895-901.

Showler A.T. (2012) Drought and arthropod pests of crops. In *Droughts: New Research*, Neves, D.F. and Sanz, J.D. (eds.), pp. 131-156. Nova Science, ISBN 978-1-62100-769-2, New York, New York, USA.

Showler A.T. and Reagan, T.E. (2012) Ecology and tactics of control for three sugarcane stalkboring species in the Western Hemisphere and Africa. In:

Sugarcane: Production, Cultivation and Uses, Concalves, J.P. and Correia, K.D. (eds.), pp. 1-14. Nova Science, ISBN 978-1-61942-214-8, New York, New York, USA.

Showler A.T., Wilson B.E. and Reagan T.E. (2012) Mexican rice borer, *Eoreuma loftini* (Dyar) (Lepidoptera: Crambidae), injury to corn greater than to sorghum and sugarcane under field conditions. *Journal of Economic Entomology* (submitted), ISSN 0022-0493.

Smith J.G. (1969) Some effects of crop background on populations of aphids and their natural enemies on brussels sprouts. *Proceedings of the Association of Applied Biology* 63: 326-330.

Smith R.W. and Whittaker J.B. (1980) The influence of habitat type on the population dynamics of *Gastrophysa viridula* DeGeer (Coleoptera: Chrysomelidae). *Journal of Animal Ecology* 49: 225-236.

Sosa O. (1990) Oviposition preference by the sugarcane borer (Lepidoptera: Pyralidae). *Journal of Economic Entomology* 83: 866-868, ISSN 0022-0493.

Steindl, D.R.L. (1957). Host range of the sugarcane stunting disease virus. *Journal of the Australian Institute of Agricultural Science* 23: 238.

Tate H.D. and Vandenberg S.R. (1939) Transmission of sugarcane mosaic by aphids. *Journal of Agricultural Research* 59: 73-79.

Taylor R.L., Maxwell B.D. and Bolk R.J. (2006) Indirect effects of herbicides on bird food resources and beneficial arthropods. *Agriculture, Ecosystems and Environment* 116: 157-164.

Todd E.H. (1964) Sugarcane mosaic on St. Augustine grass. United States Agricultural Research Service Plant Disease Report 48: 442.

Topham M. and Beardsley J.W. (1975) Influence of nectar source plants on the New Guinea weevil parasite, *Lixophaga sphenophori* (Villeneuve). *Proceedings of the Hawaii Entomological Society* 22: 145-154.

Ulloa M., Bell M.G. and Miller J.D. (1982) Losses caused by *Diatraea saccharalis* in Florida. *American Society of Sugar Cane Technologists* 1: 8-10.

Van Leerdam M.B., Johnson K.J.R. and Smith J.W. (1984) Effects of substrate physical characteristics and orientation on oviposition by *Eoreuma loftini* (Lepidoptera: Pyralidae). *Environmental Entomology* 13: 800-802.

Van Leerdam M.B., Johnson K.J.R. and Smith J.W. (1986) Ovipositional sites of *Eoreuma loftini* (Lepidoptera: Pyralidae) in sugarcane. *Environmental Entomology* 15: 75-78.

Van Zwaluenberg R.H. (1926) Insect enemies of sugarcane in western Mexico. *Journal of Economic Entomology* 19: 664-669.

Way M.J. (1977) Pest and disease status in mixed stands vs. monocultures: the relevance of ecosystem stability. In: *Origins of Pest, Parasite, Disease, and Weed Problems*, Cherrett, J.M. and Sagar, G.R. (eds.), pp. 127-138. Blackwell Scientific, ISBN 0-632-00226-3, Oxford, England.

White W.H., Reagan T.E., Smith J.W. and Salazar J.A. (2004) Refuge releases of *Cotesia flavipes* (Hymenoptera: Braconidae) into the Louisiana sugarcane ecosystem. *Environmental Entomology* 33: 627-632.

Wilcox-Lee D. and Loria R. (1987) Effects of nematode parasitism on pant-water relations. In *Vistas on Nematology*, Veecher, J.A. and Dickson, D.W. (eds.), pp. 260-266. Society of Nematologists, ISBN 978-1-62100-769-2, Hyattsville, Maryland, USA.

Wilson B. (2011) Advanced management of the Mexican rice borer (*Eoreuma loftini*) in sugarcane. M.S. thesis, Louisiana State University, Baton Rouge, Louisiana, USA.

Youm O., Browning H.W. and Gilstrap F.E. (1988) Population dynamics of stalk borers attacking corn and sorghum in the Texas Rio Grande Valley. *Southwestern Entomologist* 13: 199-204.

Yadav R.L. (1985) Improving cane yields in sugarcane and wheat companion cropping. *Indian Farming* 35: 17-21.

Zimdahl R.L. (2004) *Weed-Crop Competition: A Review*. Wiley-Blackwell, ISBN 0470290102, Ames, Iowa, USA.

INDEX

C

D

G

H

M

N

Q

R

S

W

Y

Z